A History of
Keeping and Managing
Doubled and Two-Queen Hives

Alan Wade

*To Lynne
and to our children
Andrew and Alice*

NB

Northern Bee Books

A History of Keeping and Managing Doubled and Two-Queen Hives
Copyright © Alan Wade
Canberra Region Beekeepers, 2021

Front cover photo by:
Alan Wade
Hillside Station
Narrabundah Lane
Symonston ACT Australia 2609

Back cover photo: Author photo by:
Tim Geoghagen

Published 2021 by
Northern Bee Books,
Scout Bottom Farm,
Mytholmroyd,
West Yorkshire
HX7 5JS (UK)
Tel: 01422 882751 Fax: 01422 886157
www.northernbeebooks.co.uk

ISBN 978-1-914934-16-2

Design and artwork DM Design and Print

Preface

The first ever production hives operating with more than one queen were invented one hundred and forty years ago. One hundred years later, almost to the day, fellow Canberra beekeeping club member Allan MacLean and I moved our bees to a sheltered gully on South Lanyon Station close to the tiny Hamlet of Tharwa. South Lanyon is located about fifteen kilometres south of Canberra. Allan was soon to give away beekeeping to pursue more useful and less time-consuming pursuits. So I was left to manage my sixteen or so hives – many more when I ran nuclei to support a short-term endeavour to raise queens – as best I could.

The Tharwa district is prime honey country. Spring comes with a flush of weeds growing on the fertile Murrumbidgee River floodplain, Patterson's Curse (*Echium plantagineum*), South African Cape Weed (*Arctotheca calendula*), Blackberry (exotic *Rubus* species), Crack Willow (*Salix fragilis*), Purple Top (*Verbena bonariensis*) and Turnip Weed (*Rapistrum rugosum*). So apart from annual spring requeening and judicious supering to minimise swarming I found I could leave my bees to build themselves. But the real value of the district was the eucalypt woodland comprising rich nectar bearing Blakely's Red Gum (*Eucalyptus blakelyi*), Yellow Box (*Eucalyptus melliodora*), Red Box (*Eucalyptus polyanthemos*) and late flowering Apple Box (*Eucalyptus bridgesiana*) and Mealy Bundy (*Eucalyptus nortonii*).

One other really attractive feature of the district was that were few other bees around. I was never able to find any wild honey bee colonies in any of the many hollow eucalypts surrounding the apiary though undoubtably there were a few nearby. Anyway, and for reasons I could never fathom, I never did come across other beehives in the district. This meant that overstocking was never a problem and that I did not have to worry about diseased hives nearby.

The problem for me became one of finding the time to harvest and process honey over the Christmas break.

....

To this point, I had read many of the classic books on beekeeping, Eugene Killion's *Honey in the Comb,* Harry Laidlaw's *Contemporary Queen Rearing,* Edmund Wedmore's *A Manual of Beekeeping for English-speaking Bee-keepers* and John Eckert's and Frank Shaw's *Beekeeping: Successor to Beekeeping* by Everett Phillips among more. *The Hive and the Honey Bee* was always favourite source of inspiration. In it I had read Robert Banker's account of running two-queen hives so I decided to give operating hives with an extra queen a shot.

While I harvested a tonne of honey off about eight hives over one of those summers, most coming from two-queen hives, I really knew little about how they actually worked. In truth I had simply piggybacked pairs of eight-frame double hives – with new queens – uniting them using the newspaper method and employing a queen excluder to keep the queens apart.

I soon had roaring four-decker eight-frame brood nests containing the two queens separated by an excluder. Looking back there must have been ample room for both queens to lay, sixteen frames for each queen, and clearly recall making the effort to pull out fully capped frames of honey that had accumulated in those brood chambers.

And of course, once the main flow had started, I piled on as many supers as I could lay hands on, that is above a second queen excluder. What I hadn't reckoned on were cattle in the paddock that found that tower hives made excellent back scratchers. The toppled hives were soon reassembled and strapped to study star pickets but I found I had to climb a ladder to remove ever filling supers and that I had to make regular return trips with trailer loads of extracted stickies to handle the rivers of honey.

· ·

Five years ago the Canberra Region Beekeepers formed a *Two-Queen Hive Special Interest Group*. A few staunch members, Frank Derwent, John Robinson, Dannielle Harden and myself ran club horizontal and vertical two-queen hives as well as stacked doubled hives. By then we knew a whole lot more about the workings of hives with an extra queen. Others including Agnes Koros, Danny Hunt, Harry Sydrych, Bill Scanlan and Shane Fergus lent heavy lifter skills in routine checks and honey harvesting. A happy outcome of this collaboration has been the willingness of a few club members to conduct multiple-queen hive experiments with their own bees.

Frank Derwent is a horizontal Langstroth hive enthusiast. He swaps out full depth frames from his regular Langstroth hives and to that extent this has made requeening and frame handing straightforward. About two years ago Frank built a sixty-two frame long hive designed with vertical queen excluders to operate as a two-queen hive. He is presently modifying that hive to operate as a superable four-queen colony, one that he has also designed to function as a quadrupled hive, four separate but closely juxtaposed hives, over winter.

Dannielle Harden is presently trialling a vertical doubled hive that we will come across. It will be interesting to see how it comes away in mid August, that is late in Canberra's long winter.

John Robinson, an electronics whizz, has instrumented his doubled and two-queen hives. From his living room he hopes to get a better understanding of the internal dynamics of hives operating with that extra queen.

Encouraged by Jeremy Burbidge at Northern Bee Books, I set out to pull together the short articles on doubled and two-queen hives that I had cobbled together and shared with Canberra Region Beekeepers. Some were subsequently published in *The Australasian Beekeeper* with a few recently republished in *The Irish Beekeeper* (*An Beachaire*). That brought me to Jeremy and to *A History of Keeping and Managing Doubled and Two-Queen Hives…*

Contents

Introduction

Honey bee colonies are linked inextricably to a single reigning queen. This is a condition common across all twelve known honey bee species. Yet during periods of regular queen replacement a number of potential replacement queens, so-called gynes, must be and are present. Less well known is the fact that after any transition involving supersedure both mother and daughter may continue to lay together though this condition is transitory. In practice the old queen dies or is removed and a single daughter queen assumes the sole egg-laying role. Under more aberrant conditions, such as may occur when swarms unite, a new colony may be established with two laying queens but it, too, will eventually revert to the single-queen condition.

The traditional practice of separating most of the brood from the queen to arrest the swarming impulse, the Demaree plan, may also result in the establishment of a second laying queen.

It is therefore came as a shock for beekeepers to discover that honey bee colonies could be managed with a second queen or even additional queens. Since two or more queens have the potential to build far larger populations than a single-queen colony could ever achieve on its own and since larger colonies equate to potentially much larger honey harvests the discovery of multiple queen hives has been vigorously pursued. However efforts to build such very large colonies has proved to be extremely challenging. This is the story of *A History of Keeping and Managing Doubled and Two-Queen Hives*.

....

Two types of multiple queen hives are recognised. They are distinguished by the extent of separation of their queens.

One type is built on what are effectively single-queen hives, any number of them. They share common honey-storage supers and to that extent they can be classified as multiple queen hives. Most often they are run as two abutted single-queen hives with storage space provided above a bridging excluder (Figure 1a) and schematic (Figure 2a). These have been dubbed *doubled hives*.

The other type, where there are two, three or more queens in the same hive (Figure 1b) and schematic (Figure 2b), is a more complex entity. In varying degrees the laying pattern of queens is synchronised, any significant departure from which leads to the colony reverting to its ancestral single-queen hive condition.

Figure 1 Multiple queen hives set up in author's back yard mid October 2019:
(a) doubled hive fully supered; and
(b) vertical two-queen hive also well supered.

Such hives, most often *two-queen* hives, have some defining characteristics. Well managed they are almost swarm-proof. In practice they can test the skill of the most experienced beekeeper not only in their setup but in the fact that they contain an overwhelming number of bees. Putting that together with the fact that tower colonies usually require pole support and long arms to reach the uppermost supers suggests they are not for every beekeeper.

H5 H4 H3 H2 H1 x e QA QB e e

H3 H2 H1 x ---- e QB x ---- e QA e

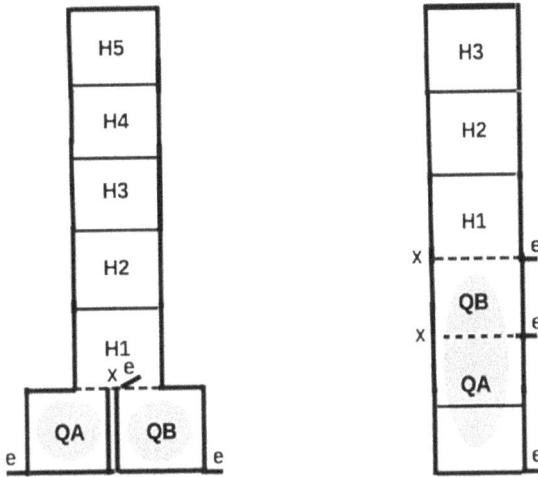

Figure 2 Schematic of double and two-queen hives shown in Figure 1:
e = entrance; x = excluder; H = honey super; QA, QB = queens:
(a) doubled hive employing independent single eight-frame brood boxes, supered above a centrally placed excluder with a riser rim and with an entrance allowing bees to access honey supers without having to traverse brood frames; and
(b) consolidated brood nest two-queen hive with tiered brood boxes, similarly supered above a second queen excluder and giving field bees direct access to honey supers. The lower brood box is extended to a second super as in practice the lower queen supports more brood than the second upper queen.

The doubled hive

In 1892, a Cottager beekeeper George Wells from Aylesford Kent announced a method of running pairs of closely abutted single-queen colonies over which were placed a 14-to-20 frame queen excluder and shallow honey supers spanning both brood boxes. These doubled hives so outperformed equivalent pairs of single-queen hives that it precipitated a Wells System craze to emulate his achievement. The journey of Wells discovery is detailed in his 1894 pamphlet the *Two Queen System of Bee Keeping*. The pamphlet has been reproduced as an appendix to this book in facsimile form.

In doubled hives the brood nests operate more or less independently, the excluder preventing queen migration to the adjacent nest. The bees work together harmoniously to harvest nectar and store honey, to share colony odour and to a limited extent queen pheromones.

Doubled hives are simple to set up and operate and are readily dismantled at the close of the honey gathering season. Wintered together – but as separate abutted colonies – they come away quickly in spring so can be manipulated to operate year round.

The two-queen hive

Then in New York Stare in 1907 E.W. Alexander made a parallel but more startling discovery. He found that one or more queens could be introduced to an already queenright hive. To do this he specially conditioned his bees to make any number of queens acceptable. He operated his multiple-queen hives excluder-free but under exceptional honey flow conditions. By the 1920s it was found that two-queen hives could be reliably operated – employing excluders – but again only under favourable conditions.

Remarkably multiple queen colonies can fully synchronise their brood raising, the most efficient arrangement being achieved where brood brood nests are juxtaposed. There the role of the queen excluder is to keep the queens apart.

Two-queen hives require constant attention and some beekeeping acumen to operate well. They must be set up each spring under exacting conditions and are best established from strong colonies. Two-queen hives are intrinsically unstable. Under conditions of dearth they have a high propensity to revert to the single-queen condition. Timing of operations, as in any good single-queen hive management program, is essential to maximise the honey crop but is more critical in the operation of multiple-queen hives. Any additional queen maintained beyond the tail of the main honey flow results in the colony consuming large amounts of stores.

This all said, two-queen hives are probably the most powerful honey gathering machines on earth though this capability may be matched by the giant Himalayan Cliff Honey Bee, *Apis laboriosa*, whose numbers can reach of the order of 100,000 individuals.

....

Multiple queen hives are operated to harness the laying power of an additional queen or queens. Queens need to be matched for their laying capacity, ideally also for provenance and age. In practice more effort is required than is needed to operate equivalent pairs of single-queen hives. However in both average and good years, this effort is more than rewarded by a spectacular honey harvest.

Doubled hives, and to a lesser extent two-queen hives, have gained a following amongst sideline beekeepers, those with the time and who recognise they can employ fewer hives to increase their honey production as well as to better pollinate their orchard and vegetables. Two-queen hives have also found a number of specialist applications, not least in building nucleus and weak colonies and in sophisticated requeening operations.

Part I examines the history of doubled, tripled and multiple single-queen hives capped with common honey supers. Part II covers the history of hives operated with multiple, usually only two, queens in the one hive.

Part I – Doubled Hive Beekeeping

Go to the ant, thou sluggard; consider her ways, and be wise:

Which having no guide, overseer, or ruler,

Provideth her meat in the summer, and gathereth her food in the harvest.

Proverbs 6:6-8

The inventor of the doubled hive, George Wells 1835-1908

I – The Discovery of the Doubled Hive

Isn't it funny
How a bear likes honey?
Buzz! Buzz! Buzz!
I wonder why he does?

Winnie The Pooh
A.A. Milne 1926

When George Wells announced a system of running hives with two queens it took the beekeeping community by storm. The background to the discovery of doubled hive beekeeping is outlined in his 1894 pamphlet[1]. Lost to the beekeeping world for over a century and *To be had of the Author only*, the Wells pamphlet, unearthed by Northern Bee Books, is republished here in facsimile form: it complements an extensive literature on doubled hives chronicled in letters written by Wells.

Doubled hives, pairs of single-queened hives sharing common honey storage space, were never thought possible until the late 19th Century. Even today finding apiarists willing to operate hives with a second queen is a distinct rarity.

Early rumblings

Prior to 1890 Wells had established an ingenious scheme for overwintering two queens on the one hive stand. To do this he divided his fourteen-frame full depth Langstroth hive bodies equally with a tight fitting central fine wire screen, each section housing a separate single-queen colony with its own entrance.

To raise spare queens Wells carefully cut out swarm cells to furnish queen stock for his doubled hives although he also employed captured swarms. These carefully paired colonies fared much better and consumed fewer stores over winter than stand alone single-queen colonies.

From these early investigations, he went on to replace the bee-proof screen division with a punched reinforced thin sheet of pine board, the so-called Wells dummy. Wells observed that bees clustered centrally on the dummy board and that they avoided propolising the interconnecting holes[2].

Wells autumn preparations paid off handsomely. At the end of winter he united the colonies by removing the central dummy board deploying spare queens to other overwintered colonies needing requeening or gifting them to local beekeepers. The resulting strong single-queen colonies were in a prime state to build bees quickly on the traditional single-queen hive plan for summer honey production.

An informed discovery

Then in 1890 Wells reserved one of his overwintered doubled hives to supply a queen to a local beekeeper. That beekeeper never turned up and Wells, faced with managing a very powerful doubled hive, further experimented by overlaying the double-brood chambers with a large queen excluder, quaintly referred to as excluder zinc. He topped this doubled brood nest with a shallow honey super to accommodate the surplus bees. He had figured that his bees, having already been acclimatised to a common colony odour, would mix without fighting above the excluder.

To his surprise, the colony proved exceptionally productive, so much so that he piled on more supers and immediately determined that he would never operate stand alone single-queen hives again. Always capable of taking considered advice, he nevertheless felt obliged to reconsider this decision when challenged about how well his hives had actually performed. He patiently trialled his doubled hives against hives headed by one queen, demonstrating conclusively the superior performance of hives with the extra queen. But to make his system work he had also to tackle the age-old problem of swarming, a task he tackled with some alacrity.

A swarm control strategy

Prior to 1890 Wells had experienced the classic problem of swarming faced by all beekeepers. As George S. Demuth noted some years later[3]:

> Gradually methods were devised for the prevention of afterswarms, and systems of management were worked out whereby the actual working force of the colony is not divided by the issuing of the prime swarm. During more recent years methods have been devised by which swarming is either prevented entirely or the act of swarming is anticipated by the beekeeper, which permits the control of swarming without constant attendance.

By supplying just enough room in the brood nest, but not too much, and by always having young queens at hand, Wells found that he could largely avert swarming and truly optimise colony buildup. To do his he adopted an elaborate scheme of filling out his brood boxes with insulation and space-filling dummy frames increasing the room needed as brood emerged carefully pacing each colony to expand and fill out its brood chamber.

These early findings, succinctly and eloquently described in the pamphlet, set the broader scene for the operation of doubled hives.

II – The Wells Doubled Hive

I will arise and go now, and go to Innisfree,
And a small cabin build there, of clay and wattles made;
Nine bean-rows will I have there, a hive for the honey-bee,
And live alone in the bee-loud glade.

The Lake Isle of Innisfree
William Butler Yeats

The aforementioned George Wells, a retired brick-maker foreman became famous not by making bricks but by harvesting astounding amounts of honey from his Aylesford, Kent back garden. His hives were the most productive in all of Merry England.

His findings dominated the pages of the prestigious *British Bee Journal, Bee-Keepers' Record and Adviser* throughout the 1890s and well into the nineteen naughties.

Prior to 1890, as Wells explains in his pamphlet, he had been working quietly on a number of backyard schemes, ones that would eventually out-fox every Cottager and gentleman beekeeper seeking to make a living out of bees in the British Isles.

Our journey of discovery into the origins of the Wells System began with the discovery of a 1904 article authored by *Bee* in the *West Gippsland Gazette*[4]. A little further delving signalled that the article was thoughtfully lifted from the *Farmer and Stockbreeder*, a rag devoted entirely to British rural agricultural life and gentlemanly pursuits. A syndicated news report it may have been, but it makes a good read. Here is an excerpt:

> *Mr Wells, by accident or design, found out that by arranging for the circulation of air through the two stocks, and by keeping the queens in their own compartments, the bees would unite and work peaceably together in a common super...*
>
> *In this way he produced immense stocks, and he was able to show returns in hive never known before in the British Isles.*

By accident or design is a misnomer. Wells was a disciplined and well organised beekeeper to whom nothing came except of diligence and a willingness to think matters through and to do things well. The first substantive details of his discovery come from the record of the Thursday 31 March 1892 *Quarterly Converzatione* meeting of The British Bee-Keepers' Association[5]. This revelatory conversation was continued at a high level at the *Quarterly Converzatione* at another meeting of the British Bee-Keepers' Association in April 1893[6].

The existence of two queens in a colony, a young queen in the process of superseding her mother was of course well known by this time, as was the notion of inducing a colony to raise a second queen by separating brood from the main brood nest[7]. The only inkling of the oncoming Wells' storm came in 1891 from a letter from a Dr G.L. Tinker[8]. This early interlocutor in Wells correspondence was a person who gave great insight into the discovery by Wells, but whose attention appears to have then been diverted by his pecuniary interests. Sadly he abandoned the Wells' cause, failing to respond to requests for his promised further correspondence, instead investing his interests in the likes of a new fangled queen excluder that he advertised in the same *British Bee Journal* in 1891.

Wells life became frenetic but his generosity of spirit and willingness to support the beekeeping fraternity made him a living legend. In perusing hundreds of letters to the *British Bee Journal and Bee-Keepers' Adviser*, I came across an April 1894 letter penned by a Helen Laurence from Leeds[9]:

> *... The best results were got from the Wells hives; the one with the driven bees of 1892 giving me over 90 lb. of honey, but no swarm. With some of that honey I took first prize at the Yorkshire Show, second at Goole, and commended at the Dairy Show, London, these being the only times I entered it for exhibition.*
>
> *In another Wells hive (fifteen frames) I placed a swarm on June 16, on full sheets of foundation, and two days later another swarm in the other compartment. As honey was coming in so fast, and I was leaving home, I at once put on the excluder and shallow frames.*
>
> *When I looked under the quilt on my return a fortnight later to see how the bees were getting on, I was astonished to find the frames full of honey, and all sealed over. Our honey returns were quite a revelation to some of the old bee-keepers here...*

Using his new strategy Wells was producing both extractable and section comb honey from early spring at a time when most successful beekeepers would have been focussed on building bees and keeping swarming in check.

George Wells the apiarist

But what of George Wells the man and his influence on beekeeping? He had seemingly retired to take up beekeeping in earnest and possibly to better support his large and extended family[10].

He was also a gentleman, a lucid and generous communicator with a great sense of self assurance and modesty, well-capable of defending himself and who kept the most meticulous records. Perhaps his working-life experience as a brick-maker, enduring some of the radiant kiln firing heat, helped him fend off his most ardent critics.

We are fortunate in having a voluminous collection of his responses to enquires from individual beekeepers most of whom were, like him, Cottager beekeepers. A close reading of this correspondence shows that he not only really understood bees but that he knew that bees well-kept would out-perform those of most beekeepers. He realised he was onto something with his discovery that running two colonies together increased honey production and determinedly set out to communicate his findings.

A report of Wells' February 1894 lecture[11] to a meeting of the Northumberland and Durham Bee-Keepers' Association typifies his willingness to share his skills and ideas:

> *Mr Wells brought with him a double-queen hive, which he took to pieces before the audience, and, while describing his arrangements, gave many valuable hints to those interested in bees. In the course of his remarks he intimated that he had experimented for results with five single-queen hives v. five double-queen hives, and found that the single hives gave 205 lb [93 kg] of honey, or an average of 41 lb [18.5 kg] each, while the double hives gave 789 lb [358 kg], or an average of 157 lb [71 kg] each.*

That beekeepers associations sponsored many such meetings is indicated by an 1894 communication[12]:

> *Mr Wells is to be congratulated upon the complete success of these pioneer meetings. The addresses were admirably delivered, and embellished with homely wit, which kept the audiences in excellent humour.*

A later visit to Wells' apiary by esteemed members of the British Beekeepers Association[13] noted that his bees were kept in immaculate condition. Wells' hives featured in several of *The apiaries of our readers* series in the *British Bee Journal* including hives in his back yard[14].

MR GEO. WELLS'S APIARY, ECCLES, NEAR AYLESFORD, KENT.

Figure 1.1   George Wells' Apiary 1899

The illustration (Figure 1.1) depicts extended shallow supers above deep double brood nests, landing boards and verandah shelters for separate brood nests and vertical separators to reduce risk of queens migrating to their adjacent compartment. Amongst the notes made by his visitors was the remarkable observation:

> *...Bee-keepers of experience will appreciate this fact because they know how seldom brood is found on the outside of outside combs in mid-April. Here, however, in every hive examined was plain evidence that the bees of both queens formed one continuous cluster, extending right through both brood chambers; the perforated divider inserted in the centre causing no break in it.*

There are literally dozens of accounts of successful operation of doubled Wells' hives of which one example may suffice. In a report of a W. Martin[15] of High Wycombe, a beekeeper who had unsuccessfully experimented with operating colonies with two queens, we hear of his discussion with another beekeeper who had had some resounding success:

> *Friend Nicholls[16], who I know very well, was at my place some time ago, and our talk as usual turning on bees, he complained of it being a bad year with him, and that he should have to feed. Yet on opening my journal I found to my surprise that he had been trying the Wells hive and had taken 183 lb. from it. And that, too, in a bad year, when some bees worked on the old plan were about starving! Then comes your correspondent, Wm Tustain, and beats friend Nicholls by 73 lb, with 256 lb. from his Wells, so even with the 10 lb. of sugar and the nice cake of candy that the hive wants for winter, it is a big take.*

Such was the demand to explain the Wells System that Wells was scheduled to speak at five meetings in one week in early 1894[17]:

> *The Committee of this Association regret that, owing to pressure of business, Mr G. Wells could not fulfil his engagement to deliver addresses in Northumberland and Durham last December. Reports from local correspondents showed that his visit was awaited with extraordinary interest, and they therefore renewed their negotiations with him, and now have pleasure in announcing that, having made special arrangements, he will address meetings as follows: February 12, 1894, Newcastle in the Mining Institute, February 13, Consett Assembly Rooms; February 14, Whittingham Schoolroom, February 15, *Cambo. February 16, *Bedlington (Station). February 17, *Riding Mill.*
>
> **These meetings are under the auspices of the Northumberland County Council. J.N. Kidd, Hon. Sec. N and DBKA*

The dutiful Honorary Secretary, J.N. Kidd[18], was the author of the article that led to this tale of rediscovery and that had found its way into the *West Gippsland Gazette*.

Advertisements for a number of Wells Hive designs[19] appeared in the *British Bee Journal* in 1893 and 1894 (Figure 1.2) though there were others[20].

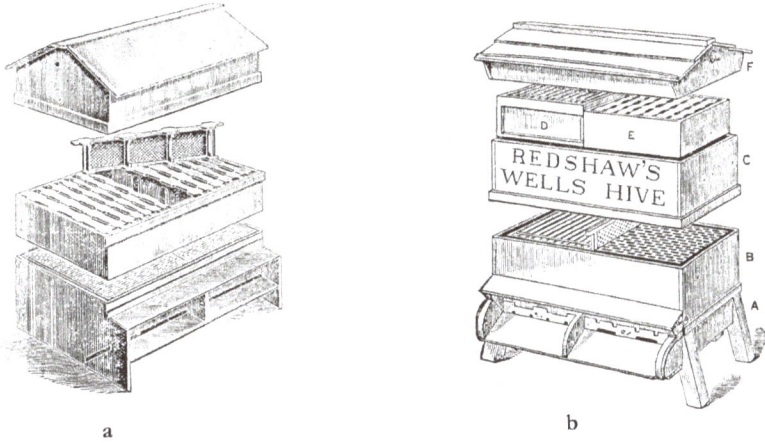

Figure 1.2   Images of Wells hives:
(a) Harrison's Wells hive No.4; and
(b) Redshaw's Wells Hive No. 6.

Success of the doubled hive scheme is also evident from other beekeepers of whom we know less but who embraced the Wells System. The editors of the journal, ever keen to report exemplary beekeeping practice, featured other Wells setups such as that found in Rymer's apiary (Figure 1.3)[21].

MR. J. RYMER'S "WELLS" APIARY, LEVISHAM, YORKS.

Figure 1.3   Rymer's apiary in Yorkshire (1898)

In the first of his annual apiary reports, Wells provides details of the relative performance of ten single-queen and five two-queen-colonies[22]. This report gave more detail than was provided in his February 1894 lecture to the meeting of the Northumberland and Durham Bee-Keepers' Association. For the individual beekeeper, still sceptical of his findings, Wells recommended the simplest of trials to compare the performance of two single and double queened hives:

> *Most bee-keepers can get single stocks strong enough to gather the main crop of honey in June and July, but not many get them up to full strength in time for the fruit-bloom and other early-flowering plants in April and May; so when any one wishes to make a fair comparison between the two systems of working, they should start, say, about the middle of October in the following way: Select four single stocks, each with young queens proved to be about equal in laying powers. Let each queen have about the same amount of brood, young bees, and plenty of food, so as to start the winter as nearly as possible on equal terms. Give them all the same amount of attention, keep a debtor's and creditor's account of all their requirements, labour, &c, included. This (or some similar plan) will show the difference between the two systems.*

An important feature of the Wells System was the already noted finding of a cluster of brood at the perforated dummy board[23] (Figure 1.4):

> *...Let us suppose that at the beginning of April there is a hive of each kind standing side by side, and containing an equal number of bees. A peep into the hives will show that the bees of the twin-hive form two clusters, one in the centre of each compartment, while the bees of the Wells hive form only one cluster, for the perforated dummy is practically no division at all.*

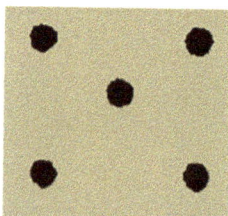

Figure 1.4   Figure-of-five punched pattern of Wells dummy[24]. Holes were ¼" (6 mm) diameter and were spaced at 1" (25 mm) intervals

This finding was to be replicated by John Hogg 100 years later that we will visit in Part II. In Hogg's two-queen system, with a similarly large single brood nest, the queens are instead separated by an excluder but formed a single large brood nest.

There were many critics of the Wells System not least sour grapes reflections by the likes of Ward[25]. Wells[26] was quick to respond to such faux criticism, here noting courteously:

> It appears to me that nothing I can write will convince Mr Ward of his errors, so I will not trouble him any further in that way. I may, however, be allowed to say that all I have in the bee line is open to his inspection at any time, and I should be most happy to explain anything or everything on the spot should he think it worth his while to visit me, where he will find the hospitality of a friend.

In another response to criticism of his system Wells[27] notes:

> In conclusion, I would say that one thing is quite certain, namely, the two-queen ball has started rolling, and I quite believe the man is not yet born who will live to see it stop.

Who was the inventor of the doubled hive?

So was George Wells the original inventor of the doubled hive? It certainly appears so and in any case he made detailed accounts of his findings partly in response to a challenges by beekeepers hoping to make the claim for themselves, to those who had deprecated the Wells System or to those who had failed to substantiate their findings[28].

Let us examine the record of the more serious claimants.

Samuel Simmins, a prominent beekeeper of the early 20th Century, provided critical commentary on the operation of such hives, a brief summary of which appeared in a 1908 edition of *Bee World*[29] and later in more detail in his 1914 book, *A modern bee-farm*[30]. In this book he noted:

> Working two or more queens in one hive.— Many years ago Dr Stroud, of Port Elizabeth, South Africa, mentioned in the British Bee Journal that he had a system of working any number of queens in one hive or colony, and that he had long practised that method.

He added preciently:

No details, however, were given by him, and if there were any merits in his process he appears to have been careful to keep the supposed secret to himself.

Dr James Stroud was an authority on African bees[31] but appears likely to have worked mainly with the Cape Honey Bees (*Apis mellifera capensis*), a bee now well-known for raising queens from laying workers and known to tolerate two queens in the one hive. So he may have simply recorded the presence of two queens in some production hives. There is certainly no reference to his operating hives with two queen hives – as claimed by Simmins – in all back issues of the *British Bee Journal*.

More significantly Stroud is recorded as having made a number of attempts to market and export South African queen bees to England during 1885-1886[32]. He records three kinds of bees, the yellow, the blackish-brown, and the mixed. He was presumably referring to what we now know as the parasitising Cape Honey Bee (*Apis mellifera capensis*) and the problematic African Honey Bee (*Apis mellifera scutellata*). It seems likely that these importations to the United Kingdom failed as their having successfully done so would likely have changed the course of western beekeeping. No one now wants the killer *scutellata* bee.

Simmins makes similar reference to running the so-called two-queen hives to a Mr Heddon who had stated[33].

... as I find that the Bees will tolerate two queens, or cells and queens, in the same hive, at all times in the season, if divided by these queen excluding Honey Boards.

Of this Simmons surmised:

Mr Heddon, of Dowagiac, Michigan, claims to have been the first to point out the possibility of working more than one queen in a hive. Doolittle and others made some practical demonstration of the fact, but neither of them preceded Dr Stroud.

Overall we see a growing recognition that a colony would tolerate two queens if separated by a queen excluder. But this was a far cry from the Wells System of employing two queens in separate brood boxes to produce bumper honey harvests. And while James Heddon claimed to have been the first to point out the possibility of working more than one queen in a hive, he – like Stroud – was either too coy to say he had exploited their honey producing potential or was more likely focussed on their use in queen raising operations:

I am quite positive that they, queen excluders, are going to aid us greatly in queen rearing, as I find that the bees will tolerate two queens, or cells and queens, in the same hive, at all times in the season, if divided by these queen excluding honey boards.

So where does this leave Wells?

There appear to be no record of Simmins claims of Heddon's and Stroud's pre-1900 findings of production hives with two queens in all the volumes of the *British Bee Journal and Bee-Keepers' Adviser* since its inception in 1873 as cited in his 1893 edition of his book.

Like Tinker, Simmons – though an eminent beekeeper – appears to have had as much an eye to the commercial prospects of the beekeeping trade (Figure 1.5) as to progressive beekeeping practice.

Nevertheless Simmins, in discussing multiply-queened hives, was close on Wells' trail:

In my 1893 edition of A modern bee-farm, I illustrated a method of tiering up single stock chambers, with two, four, or more queens, before adding the supers, and showed how to unite them safely. This plan, however, is not in any sense equal to the lateral turn-over plan, which (with my intermediate way) crowds the whole maturing working force into one stock chamber, so that the supers are more rapidly filled, and no stores can accumulate in surplus stock chambers, as in the usual tiering up methods.

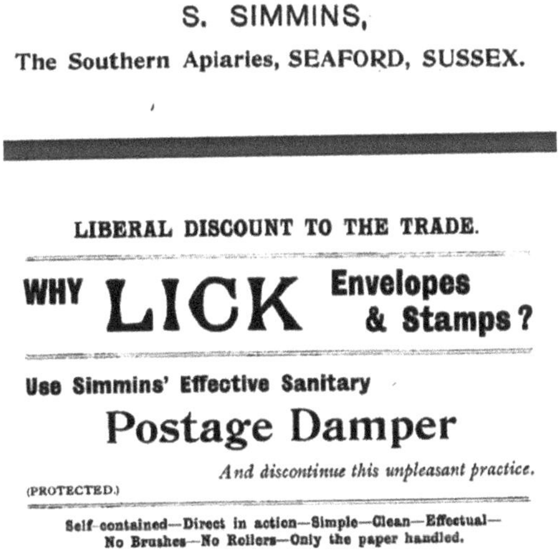

S. SIMMINS,
The Southern Apiaries, SEAFORD, SUSSEX.

Figure 1.5   Simmins mailing accoutrement for the busy beekeeper

Indeed, in that earlier 1893 edition of his book, Simmins gives a full account of the Wells System with full working drawings and provides a blueprint for their operation[34].

There are innumerable earlier accounts of queens living harmoniously together but seemingly never to create powerful honey gathering colonies. For example Alfred Neighbour's 1877 classic *The apiary, or, bees, beehives and bee culture*[35] quotes the illustrious apiarist and German editor of *Bienenzeitung*, F.W. Vogel:

> *The workers do not always decide: the matter in such case; it is, indeed, nothing uncommon, says Vogel, for two fruitful queens to be allowed to live together; and we have had instances of the same kind ourselves, without being able to give a reason other than that the exceptions prove the rule.*

Charles Dadant made a similar 1886 observation in noting the common presence of two queens in a colony where the old queen is being superseded[36]:

> *...frequently enough, two queens, the mother and her daughter, are found living in peace in the same colony, both laying and living side by side without interfering with each other. This takes place when the bees, having recognised the failing fecundity of the mother, have made preparations to replace her by rearing one of her daughters. But this abnormal state of two queens in one hive generally only lasts a few weeks, or at most a few months.*

Despite these spasmodic reports, it appears that Wells was the first person to reduce the matter of operating hives with an extra queen to a practical system of maximising honey production distilled succinctly as:

> *The stock hive is divided by a perforated wood dummy, while the bees from both sides have common access to the supers placed over excluder zinc.*

Indeed the late 1914 edition of Simmins book *A modern bee-farm and its economic management*, referring to Wells' 1892 discovery, he spilled the beans in noting that:

> *Mr Wells, of Aylesford, however, was the first to reduce the matter to practical working as a system in honey production.*

The prize for the being the first to manage honey-producing hives with two queens goes, indisputably, to George Wells.

The legacy of the Wells doubled hive

Wells success drew praise and support from the wily and astute editors of the *British Bee Journal*. Despite his many detractors, and the carping of many ignorant and incompetent bee owners, Wells fame grew and, as we have seen, he became a favourite speaker at beekeeper association meetings. On one occasion he received ninety letters of enquiry in his postbox in a single day[37]. This is not a bad score given that social media was only to come to the fore nearly a century and a half later.

The many details of operating the Wells System doubled hive are chronicled in around four hundred letters that appeared in the *British Bee Journal* between 1892 and 1909. Included amongst these were detailed annual production records for each of his Aylesford apiary doubled hives for nine consecutive years (1892-1900)[38].

As we have noted the original 1880s Wells System of hive doubling relied on a range of subtle measures including use of productive queens, dividing really strong hives in autumn (or retaining doubled hives operated over the past season), closely abutting these single-queen units using a thin-walled partition to enable each colony to share its neighbour's warmth over winter, then adopting a very proactive approach to swarm control and supering as needed. Wells surmised these findings in 1893 by stating[39]:

> *I am very pleased to find so many bee-keepers giving the double-queen system a trial, and also glad to note that some are comparing results, viz., one hive with two queens against two hives with but one queen in each; but I hope that all who are testing the double queen system on these lines will bear in mind that they do not really prove its merits or demerits unless they began their operations when packing up their bees for winter.*

Wells colonies gained attention not only because they overwintered exceptionally well but also because his annual honey crops were nothing short of spectacular. However the carpentry skills required to build a Wells hive – or indeed to partition large (typically 14-20 frame) brood chambers – now seems rather too daunting a task for even the most dedicated beekeeper, but one needed to exactly replicate his original scheme of operation.

By the 1920s the Wells System of beekeeping was falling out of favour. This was triggered by a combination of beekeepers not being skilled enough to operate colonies with more than one queen, by the labour intensive nature of the system and by its requirement for specialised hive gear that, by its nature, was too cumbersome to transport.

However the lure of overwintering strong colonies headed by young queens in doubled hives to supply a surfeit of queens and to give colonies a head start at the end of winter was never lost.

The opportunities afforded by the Wells System presents the next phase of discovery for the wider adoption of doubled hives.

III – Doubled Hives Systems post Wells

Like trains of cars on tracks of plush
I hear the level bee:
A jar across the flowers goes,
Their velvet masonry.

The Bee
Emily Dickinson

In paying tribute to the Wells System, it is important to ask which of the valuable features of the scheme have been adapted to contemporary beekeeping practice. Firstly we need to establish an understanding of how bees operate in a hive when more than one queen is present. What is the doubled hive, how does it differ from the two-queen hive – a hive with two queens in the one colony – and why are doubled hive and two-queen hives particularly effective gatherers of honey?

Doubled hives versus two-queen hives

As outlined in the introduction there are two distinct types of hive that support at least one extra queen:

- the doubled hive that comprises a pair of single-queen hives. Once well established the colonies are united by straddling an excluder and honey supers over both single-queen hives where the bees share storage space. The Wells hive is just one — and the original — version; and

- the two-queen hive[40] that has two queens in the one colony, ideally in the one broodnest. Both queens are in the same hive kept apart only by an excluder (or excluders).

The original designs in both schemes emphasised the importance of keeping brood nests well apart, in the case of two-queen hives using two excluders and sandwiching honey supers between. This notion was later discarded, the two queens in the two-queen system being found to be better housed in adjacent boxes forming a single consolidated brood nest (CBN). The multi-queen hive is explored in more detail in Part II.

There are many shared features of the doubled hive and two-queen systems. All things being equal, both exploit the power of two laying queens, the combined workforce being far more efficient than two average strong single-queen hives in gathering and storing honey. Clayton Farrar[41] showed – as we shall also see – that, as the number of bees in a colony increases, so too does its honey gathering potential. Doubling hive strength far more than doubles single-queen colony honey production, powerful colonies devoting a much smaller proportion of their workforce to tending brood. This frees up the vast majority of bees to harvest nectar and to process and store honey.

We now turn to new schemes for overwintering queens in doubled hives that emerged after Wells exited centre stage.

The Ferris doubled hive – 1906

A key feature of the Wells System is autumn preparation of doubled hives that contain young queens. Colonies come away earlier and in better condition in early spring than do conventional single-queen colonies.

Ferris[42] developed a vertically tiered brood nest arrangement for overwintering of co-located pairs of six-frame single-queen hives, misnamed two-queen colonies, in standard hive bodies. Each hive body was divided centrally by a thick partition board (Figure 1.6). In the Ferris system, the shared warmth alone resulted in more efficient use of stores facilitating late-winter, early-spring buildup.

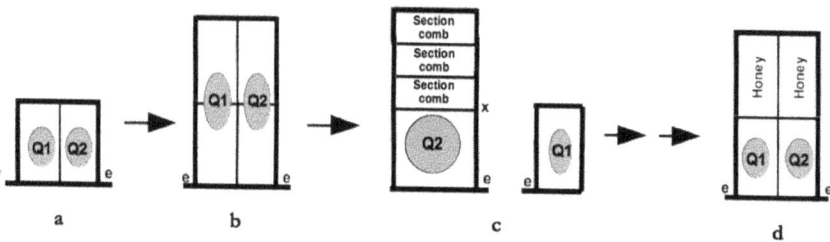

Figure 1.6 Ferris double-six system for building bees for honey flows: e=entrance; x = excluder; Q = queen:

(a) overwintered double six-frame nucleus colony is housed in a 14-frame brood box with a thick tight-fitting central division board and spaces completely filled out with follower boards on the outside walls;

(b) each colony is built to strong double six-frame singles to prepare for the honey flow;

(c) colonies are united and supered for honey flows with one queen offset to a nucleus; and

(d) in autumn, colony is split and overwintered as a double colony ready for early spring buildup (a) or alternatively wintered as a doubled tiered hive (b).

24

At the commencement of the flow, Ferris first offset queens (either one or both to inhibit swarming) to small nucleus colonies, then united the bees removing the central bee-proof division to form a double strength colony. This consolidated most of the brood and the majority of the bees from both colonies to a single full-width brood box that was then supered for the flow.

The honey supers are depicted as shallows, fitted out with section combs, as this was both common practice at the time and reflected the premium price paid for section comb honey. Ferris used weaker colonies to produce extractable honey. So while two queens were employed to build abutted colonies over winter and in spring, for ease of operation he employed powerful single-queen colonies for the actual flow. In essence the Ferris doubled hive is simply a variant of the conventional split employed to prevent swarming and to build strong colonies. Having united the colonies for the flow, at the end of the season the hive was re-split to reform the double nucleus setup.

Ferris provided detailed instructions for setting up the split chamber system – all outside walls were 1½ inch (37 mm) thick so the two colonies were well insulated. From years of experience I can attest to the importance of ensuring that dividing boards in doubled nuclei are very close fitting. This is essential to avoid queen crossover. Ferris achieved this by using very sturdy but not overly thick ½ inch (12 mm) division boards that slotted snugly into a 9/16th inch (14 mm) slots routed into the end walls of fourteen-frame boxes.

The Sladen doubled hive – 1918-1921

While the principles for operating doubled hives all have a common theme of abutting hives to enable them to overwinter well, their use to build bees for the honey flow varied considerably. For example, the inventive Sladen[43] refined the doubling scheme to forestall excessive swarming peculiar to the local Ottawa region of Ontario.

To do this he first divided his overwintered 2x5-frame doubles by transferring the stronger colony to an adjacent 10-frame brood box then removing the double screen division. Sladen then expanded both colonies on the spring dandelion flow and, to control excessive late swarming, he dequeened and re-split both hives allowing both hives to raise emergency queens to reestablish the doubled hive setup condition (Figure 1.7). Once established both these spring doubled hives were supered in much the same way as Wells had done for honey flows. This spring doubling, effecting a break in the brood cycle, was about the only reliable measure Sladen found he could take to prevent swarming.

In reiterating Sladen's findings Iona Fowls[44] reports that his spring doubled hive produced 480 pounds (218 kg) of honey while the average yield of other single-queen hives in the apiary, those where swarming was likely well controlled, was 176 pounds (80 kg).

The other spinoff from Sladen's scheme was a mechanism for making increase and producing a surfeit of young queens well into spring.

While Sladen supered above the abutted doubled hive over a common excluder – as did Wells – he united bees just prior to the flow transferring brood to one large super to inhibit the tendency of bees to store honey in the brood nest. The adaptations to different buildup and flow conditions signals the need to read local climatic and seasonal conditions to maximise the benefit of schemes employing two queens. His story is embedded in Figure 1.7 and a detailed reading of his account.

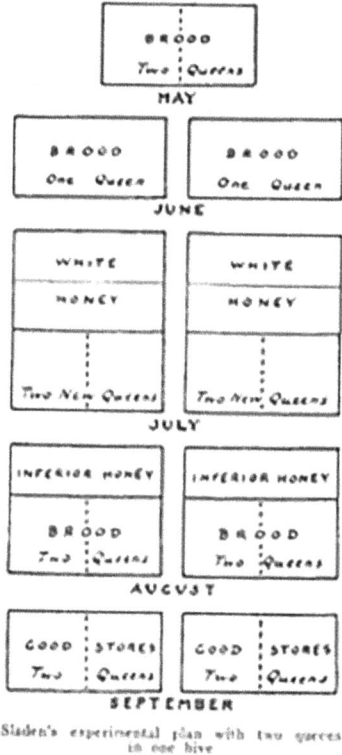

Figure 1.7 Sladen's scheme to control swarming[45]

The Atkinson dual queen, double-six system – 1920-1922

By 1920, Atkinson had not only already trumpeted his Masheath hives (Figure 1.8)[46] but was also promoting a tiered system for building bees rather than relying on lateral expansion in large brood nests[47]. Following a close reading of this early report on the advantages of tiering hives and later running them as doubled hives to overwinter bees (and to build bees for the honey flow), we may recognise his contribution as being one of very effective operation of doubled hives.

Figure 1.8   Masheath bee hive advertisements

Atkinson proposed a novel way to build bees for the honey heather flow[48] titling his system *The dual-queen work—the double-six system* made in response to Ellis's[49] urgings to expand further on his (Atkinson's) tiered system of building bees. Ellis had noted that:

> *The inventive genius of Atkinson has made this possible by the simple expedient of using two five-frame stock boxes, one above the other, in place of the usual ten-frame brood chamber, reverting to the latter during the actual honey flow only.*

Personally, I consider this principle of adjusting the depth and shape of brood chambers is a most valuable feature, and hope Mr Atkinson will let us have some luminous contributions on the subject.

Akinson, ever the entrepreneur, was never shy to promote his wares and ideas. In announcing his doubled hive scheme he stated rather disingenuously[50]:

In the working out of the system and the handling of the colonies there is a wealth of detail I wouldn't dare to expect the editors to find space for.

To get over the difficulty readers can help themselves by taking advantage of the offer in the advertisement column of the loan of typescript copies of my Cambridge lecture (1920), which goes fully into the dual-colony system and its hive.

He promptly advertised his 1920 – now lost – lecture notes[51], noting that they would detail his method of operation:

Double-six system, dual-colony working. What it is, and how effected. Typescript copies (24 pages) of Cambridge Lecture, 1920, loaned, returnable within one week, 1s., from the author, M. Atkinson, Fakenham.

Atkinson had, however, clearly argued that two queens should be employed in the building phase and that this phase should be delineated from the honey production where the strength of the colony alone matters.

Much of the drive to run two queens to build bees came from the lure of the heather (*Calluna vulgaris*) flow first championed by Ellis and Medicus[52]. Ellis, in particular was instrumental in further refining management of bees for the late and valuable heather flow[53]:

As regards dual-queen working at the moors, my initial experiment last season was quite satisfactory, and each queen kept her five combs full of brood, so what little heather honey there was all went into the sections.

The only apparent drawback was that the incoming foragers were inclined to drift largely towards one entrance of the dual-queen hives.

In broad outline Atkinson[54] concluded:

But while the Double-Six hive is perfectly equipped for this dual-queen work, already pointed out, its chief aim is to obliterate dual-queen or dual-colony work on the eve of the honey flow, and reconstitute the entire contents of the hitherto double-colony hive, as a one colony, one queen, one brood chamber (ten or twelve frames), one set of supers, and one combined force of foragers.

It means that though you have a double force of adult and active workers, only one brood chamber and queen has to be provided for out of the incoming surplus. To that extent it is superior to any dual-queen, supered force. The Double-Six system requires, and provides for, dual-colony work for bee production, the first necessity in any case. That in turn provides the material for the augmented single colony for the next stage in the season's work, viz., honey production, which in the end is what everybody with fully-fledged bee forces is after.

In the same communication Ellis[55] noted:

It is not generally known that an ordinary hive of the WBC type is quite adaptable to the dual-queen system and can easily accommodate two ten-frame colonies, each with its own queen, and separate entrance.

In a final tête-à-tête with Atkinson, Ellis[56] sought to clarify his attempts to make use of his, Atkinson's doubled hive setup:

Mr Atkinson has misunderstood me. My experiment at the heather was with two five-frame lots side by side in one hive, but supered separately, not working over (an) excluder in a joint super. The Atkinson methods have distinct possibilities, and seem well adapted to meet the needs of those who wish to secure large crops of honey by intensive rather than extensive bee-keeping.

Ellis brought the double-six system into his repertoire for operating his bees on the heather[57] for building bees:

This applies particularly to heather districts, where in the mutual interests of bees and bee-keepers it is advisable to reverse the usual procedure of clover-filled sections, and brood combs, blocked with heather honey. The latter familiar problem is possibly soluble along the lines of dual-queen working in the vertically divided Atkinson brood chambers, and experiments here last August showed the value of a supplementary queen in reducing brood-nest storage to a matter of ounces.

However in providing this detail Ellis unsuccessfully sought more detail from Atkinson:

Perhaps Mr Atkinson might give us an article on the dual-colony and double-six methods.

In a departing communication Ellis[58] further surmised his experiences using doubled hives for working the heather flow:

> *Personally I favour dual-queen working, not in the unwieldy Wells hive, but in the ordinary WBC type fitted with Atkinson vertically-divided brood chambers. A colony can be split in two after the honey flow, both lots wintered under the one roof, and, with two queens at work, it is quite possible to have thirty standard combs of brood in one hive previous to the white clover flow. For section-honey production, minus swarming, the combined working forces could be started anew on ten-frame foundation, and with a young queen just as the honey flow began. The original hive, still with its two queens in their respective nurseries, being, of course, moved aside to produce another large force of gatherers for a later honey flow, such as heather.*

In a late and prescient note to American beekeepers[59], Ellis also noted:

> *While a brood-chamber can not be too large before the honey flow, this virtue becomes a defect if persisted in after the section supers are on...*

> *As a means to this end, the Atkinson dual-queen system is attracting some attention here. On this principle two queens are wintered under one roof, each in her own six-frame brood box and by the addition of other similar boxes in spring the respective brood chambers can be expanded to any size required. Finally, both lots are combined with one queen on ten standard-sized brood-combs, when the days come for honey-gathering.*

A generalised scheme for operating doubled hives

We can distil the collective findings of Wells, Ferris, Sladen, Atkinson and Ellis into a very effective generalised scheme for operating doubled hives, for overwintering, for spring buildup and for the honey flow.

Figure 1.9 depicts the rather confusing array of doubled hive setups developed in the late 19th and early 20th Centuries in their overwintering configuration. Many much later schemes, stretching through to to the early 21st Century, were to employ oversized Langstroth hive bodies divided in an array of ways to both overwinter queens and to produce large honey crops. For example both Burkhardt and Vitez[60], and more recently Cuello[61], employed a double Langstroth hive body divided by a central screen (as had Sladen) and the prototype Wells hive. Earlier Gerstner[62] employed a hybrid horizontal doubled hive setup with brood nests on the wings, a partially gapped central unit bounded by vertical queen excluders and a centrally tiered honey super stack.

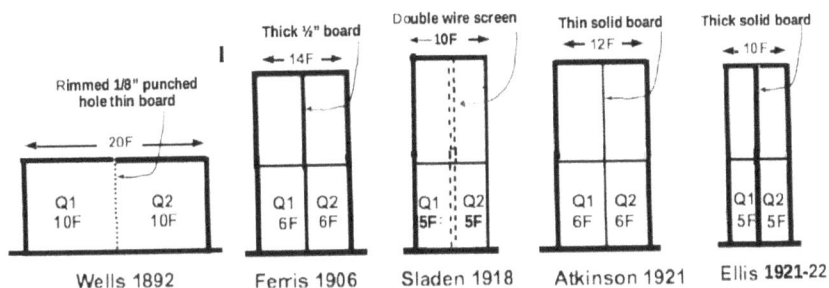

Figure 1.9   Historical doubled hive systems: Q1, Q2 = queens

Quite recently Gutierrez and Rebolledo[63] compared the productivity of tiered two-queen hives, doubled hives and traditional single-queen systems in Ranquilco, Chile. All hives with two queens outperformed equivalent pairs of single-queen hives.

Doubled hive overwintering of bees   In all approaches pairs of fledgling colonies were accommodated in the one large super, that is after splitting a colony by slipping in a bee-proof division board and introducing a caged queen to the queenless unit or requeening both units. Thus set up, the fledgling colonies shared the warmth of their neighbours' brood nest (Figure 1.10a → 1.10b) greatly improving their overwintering capacity and greatly accelerating their early spring development.

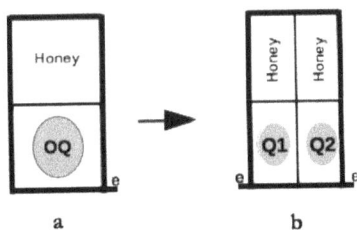

Figure 1.10   Optimised doubled hive for overwintering queens on one hive stand: e = entrance; OQ = old queen; Q1, Q2 = queens:
(a) single ten or twelve-frame colony with old queen (OQ); and
(b) hive spilt and reorganised with central tight-fitting division board with new queens (Q1 and Q2) – or an old and one new queen – introduced, with stores moved above.

Alternative setups employing tiered divided six-frame and eight-frame gear operating on the same principle would seem equally applicable. Indeed Canberra Region Beekeepers have frequently successfully overwintered bees in highly insulated Paradise high density polystyrene six-frame gear either as single 6-frame or as divided 2x3-frame colonies.

31

Doubled hive spring expansion of bees By late winter brood and bees move up into their respective upper double chambers as strong five or six-frame doubled nucleus colonies (Figure 1.11a) shown here with the empty lower double brood chamber removed. By placing this now empty twin chamber on top of this unit bees will build quickly to strong pairs of 2x6-frame (or 2x5-frame) colonies (Figure 1.11b).

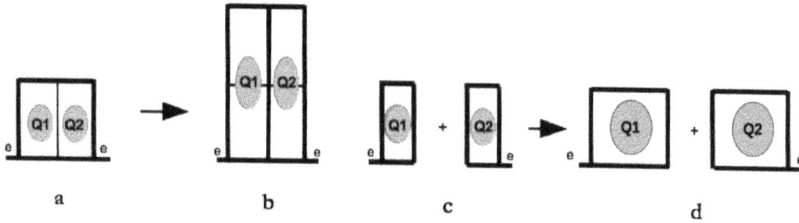

Figure 1.11   Spring expansion of nuclei in doubled hive scheme versus traditional stand alone buildup of pairs of small colonies in standard gear: e = entrance; Q1, Q2 = queens:

(a) starting double-six hive;

(b) abutted two by six-frame colonies;

(c) conventional separate six-frame nuclei; and

(d) nuclei expanded to standard twelve-frame brood chambers.

This replicates the normal spring practice of building stand alone nucleus colonies (Figure 1.11c) to full strength colonies (Figure 1.11d). The advantage gained in running tiered rather than conventional stand alone hives is that colonies build far more quickly.

Doubled hive consolidation of bees for the honey flow While the doubled hive system can be maintained to advantage through long honey flows, the general consensus is that uniting bees to form very powerful single-queen colonies and offsetting spare queens to nuclei – or using those queens to requeen colonies with old or failing queens – is simpler (Figure 1.12), easier indeed than maintaining the Wells-like doubled hive arrangement year round. In most respects it follows the earlier schemes.

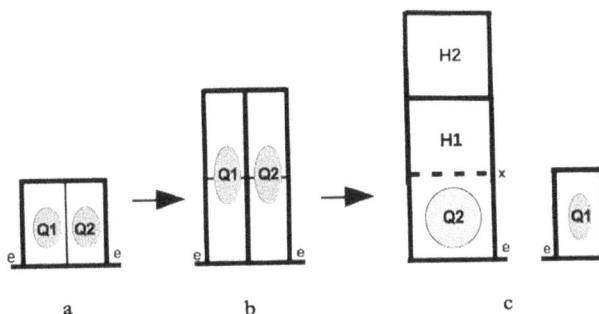

Figure 1.12   Doubled hive scheme for honey production: e = entrance; x = excluder;
Q1, Q2 = queens;
H = honey supers:
(a) overwintered doubled colony in early spring;
(b) colonies built to full ten or twelve-frame condition prior to main flow; and
(c) colonies combined as a single-queen hive with sealed brood, and most of the bees are supered for the flow.

There are many salient features of doubled hive systems that can be incorporated into routine single-queen hive management that will improve colony performance:

- requeening hives in autumn to give bees an early start in late winter and to make hives less swarm prone in early spring;

- overwintering pairs of hives from splits made up in autumn from strong colonies housing them either in large brood boxes with a central partition or in stacked brood boxes divided by a thin board or by a vertical double screen[64];

- uniting colonies in early spring – supering and feeding them as needed – deploying spare queens to requeen colonies in need of requeening; and

- further building these strong colonies quickly to better enable them to take advantage of early blossom flows and to pollinate fruit trees.

IV – Doubled Hive Management

And still more, later flowers for the bees,
Until they think warm days will never cease,
For summer has o'er-brimm'd their clammy cells.

To Autumn

John Keats

In more recent times Ray Nabors and Bill Hesbach[65] as well as Jones[66] have popularised a novel cut-down version of the original Wells System (Figure 1.13). Instead of one large brood box, divided centrally to accommodate two overwintering colonies, Nabors and Hesbach simply abutted strong single-queen hives supering them over a central excluder with full depth supers. Gouget's pyramid hive (Figure 1.13c) is an extension of this principle. Modern setups employ only standard gear, lend structural stability to the whole system and provide side-by-side access to the separate brood chambers.

However, while the shared honey supering allows conventional crop removal and standard honey supering, neither brood nest benefits from the warmth generated by a sister brood nest nor from effective sharing of queen pheromones. The potential for superior overwintering achieved by Wells, Ferris, Sladen, Atkinson and Ellis is also lost. The modern doubled hive is nevertheless a very productive unit and will out-perform equivalent pairs of single hives but only if an early start is made in spring and only if autumn preparations are made.

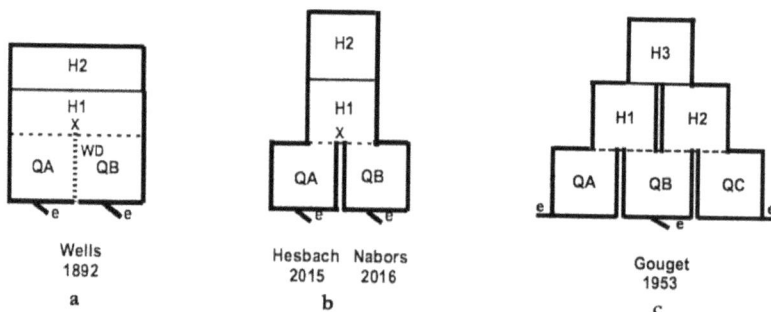

Figure 1.13   Setups with two queens in separate colonies comparing original and modern versions of the doubled hive: e = entrance; x=excluder; WD=Wells dummy; H1, H2, H3 = honey supers; QA, QB, QC = queens:
(a) original Wells System doubled hive;

(b) modern Hesbach – Nabors tower doubled hive; and

(c) Gouget's pyramid hive.

One of the most interesting inventions of the early 1950s was that of the pyramid tripled hive (Figure 1.14; schematic Figure 1.13c) formulated by Gouget[67]. Like the Hesbach-Nabors doubled hive it employed standard gear only apart from a modified bottom board and side covers. In Gouget's system, there are three brood chambers located side by side on a horizontal platform, each containing a young vigorous queen. Queen excluders are placed over each brood chamber, two supers are placed centrally over the brood chambers and a further super is placed on top of the stack to form a very stable pyramid.

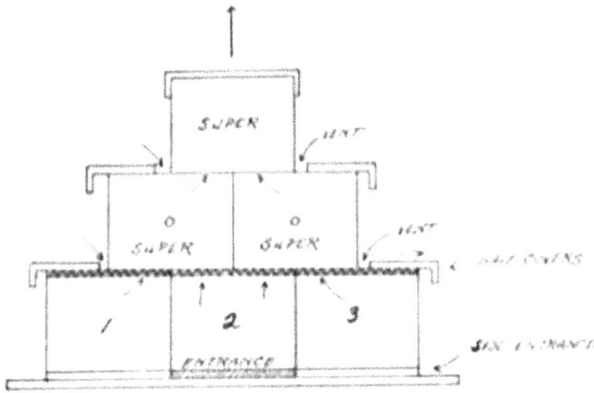

Figure 1.14   Gouget's tripled hive showing side-by-side brood chambers overlaid by standard queen excluders supered in a flexible manner using the same sized boxes: top arrow probably signals potential for further supering

An exceptionally strong startup colony would have been required if the colony were to have been split three ways to establish the tripled queen hive though other brood and bees could have been added. Gouget provides clear instructions on the operation of the pyramid hive. Notably he emphasises facility to move supers around so they fill more evenly and so that brood boxes can be inspected routinely without the need to offset honey supers (Figure 1.15). All in all the pyramid hive is a simple, yet potentially extremely powerful, colony setup.

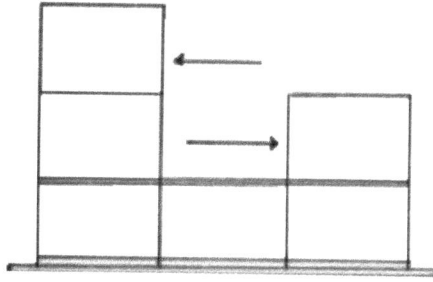

Figure 1.15 Gouget's scheme for equalising super filling during the flow and to facilitate inspection of brood boxes during a honey flow

Nabors and Hesbach describe many practical doubled hive setup measures, for example providing design details for non standard side-covers and the use of a small strip of thin masonite or linoleum to bridge the joint between the brood boxes to avert accidental migration of queens between brood chambers. They also discussed the merits and drawbacks of the horizontal system, particularly noting that honey supers can be managed in much the same way as they would be in an overly strong single-queen colony. A key advantage of the doubled hive is that each brood box can be periodically inspected to check that each is performing satisfactorily and is not honey bound. While the double hive system employs standard hive gear throughout Hesbach sensibly screws the brood boxes together to lend lateral stability to the whole system.

The honey crops obtained by doubling hives, that is using separate hives each containing their own queen, but using their combined workforce to harvest honey, are rarely achieved with equivalent pairs of single-queen hives. In the strictest sense doubled hives are not two-queen hives, well not in the modern sense that two queens can be worked together in a single colony. Nevertheless doubled hives were the first ever colonies to be managed with two queens.

The origins of the modern doubled hive remain obscure and Nabors and Hesbach were certainly not the first to use the system they promoted. Eva Crane[68] cites hives with six queens in a commercial operation in Western Australia:

> I saw multiple-queen hives in commercial use in Western Australia in 1967, that extended horizontally, instead of vertically as skyscrapers do; they were called coffin hives. The unit consisted of a row of six 8-frame Langstroth brood boxes, separated by dividers, their flight entrances facing alternately to either side of the row. A single super (honey-storage chamber), holding 50 frames,

extended across all six brood boxes, each of which had its own queen excluder. Alternatively a three-decker outfit was used, with a row of six separate honey supers below the 50-frame super...

The 1967 multiple-queen hives referred to by Crane were operated by Ken Gray in Western Australian Wandoo (*Eucalyptus wandoo*) woodland and by Sid Murdoch at Manjimup (Figure 1.16), the broad details of operation of which are outlined in Francis Smith's booklet *The Hive*.

Figure 1.16 Ken Gray's and Sid Murdoch's operations employing hives with six queens

Images: Eva Crane Foundation

Dave Flanagan has recounted his 1990 and 1991 experiences of working with a similarly extraordinary beekeeping enterprise (Figure 1.17) operating out of the tiny northwestern New South Wales town of Wanaaring on the Paroo River . He worked for Stringy Hughston who was running 2400 tripled hives.

Figure 1.17 Dave Flanagan transporting coffin hives on the Paroo River New South Wales in 1990
Photo: Dave Flanagan

The river is an oasis in a desert. It is valued not only for its Ramsar listed Paroo River Wetlands but also for its being a beekeeping mecca. The Paroo climate is however harsh a theme reflected strongly in Edward Harrington's poem *My old black billy*:

I have carried my swag on the parched Paroo.
Where water is scarce and the houses few.
On many a track on the great outback.
Where the heat would drive you silly.
I have carried my sensible indispensable
Old black billy.

The Paroo River rises in the Warrego Range west of Charleville in Queensland. It courses its way south through arid Channel Country eventually joining the Darling River in turn a tributary of the Murray River.

The semi-desert flora of the Paroo district includes Budda Bush (*Eremophilla mitchelli*) a nectar producing plant and Hop Bush (*Dodonaea* species) an important pollen producing taxon. The character of the riverine floodplain could not be more different. The Paroo River water holes and lagoons, and a shallow groundwater system, support a highly diverse and nectar-rich flora: spring and summer flowering River Red Gum (*Eucalyptus camaldulensis*), Coolabah (*Eucalyptus coolabah*), Bimble Box (*Eucalyptus populnea*), Black Box (*Eucalyptus largiflorens*) and Leopardwood (*Flindersia maculosa*) as well as winter and spring flowering flowering Napunyah (*Eucalyptus orchroploia*) – the key honey producing species – and Whitewood (*Atalaya hemiglauca*).

Hive setup. Hive establishment was based on an intense program of splitting, requeening and building hives with a heavy emphasis on culling colonies showing any sign of disease. Bees from strong hives (Figure 1.18a) were first shaken down into an empty ten frame super placed on the parent hive stand then filled out with mainly empty combs, brood and stores being transferred to an offset super on a separate stand. This super was then placed back on the hive above a queen excluder allowing bees to filter up through the screen but leaving the queen below (Figure 1.18b).

Once the bees had settled, brood, bees and stores were distributed to stand alone ten frame nucleus colonies (Figure 1.18c) and each was supplied with a purchased caged queen. These fledgling colonies were bolstered by pollen trapped from healthy hives and purchased irradiated pollen and by stores from the parent hive. Once well established the new hives were trucked to production locations on the Paroo floodplain dotted ninety kilometres upstream and downstream of Wanaaring while the old hive was retained in a home nursery.

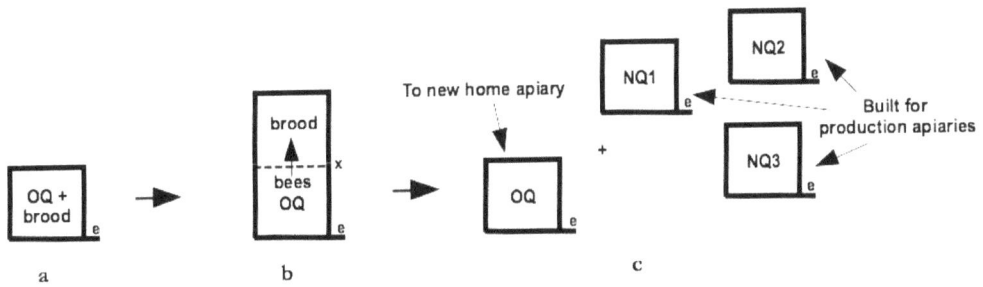

Figure 1.18 Single hive establishment: e = entrance, x = excluder, OQ = old queen, NQ = new queen:
(a) a strong colony was built;
(b) bees were shaken into an empty super on the parent hive stand and brood and stores were moved above an excluder to isolate the queen in the lower chamber; and
(c) after bees had moved up through the excluder, bees brood and stores were distributed to new brood boxes, a caged queen being introduced to each of these splits. All colonies were supplementary fed and rebuilt.

Hive operation. Apiaries, each comprising 120 pallets, were located close to permanent water holes and lagoons on the floodplain or, in the absence of permanent water, supplied a bee-friendly watering station fed by a 200 L drum.

The single brood chambers transported to production apiaries were unloaded and assembled as tripled hives (Figure 1.19a), two units to the pallet. Alternatively single chamber hives were used in an established hive requeening program where brood boxes with old or failed queens queens were removed and new brood chambers papered in.

The three-hive units were united by papering on thirty five-frame full-depth (coffin) boxes, above excluders, to form tripled hives (Figure 1.19b). These hives were dynamically operated, honey being either removed and the hives re-supered or further supered with one or more additional coffin chambers (Figure 1.19c): hot spots, sections of the large coffin supers that were filling quickly, could be further supered with a single ten-frame super (H3, Figure 1.19c) to maximise honey storage.

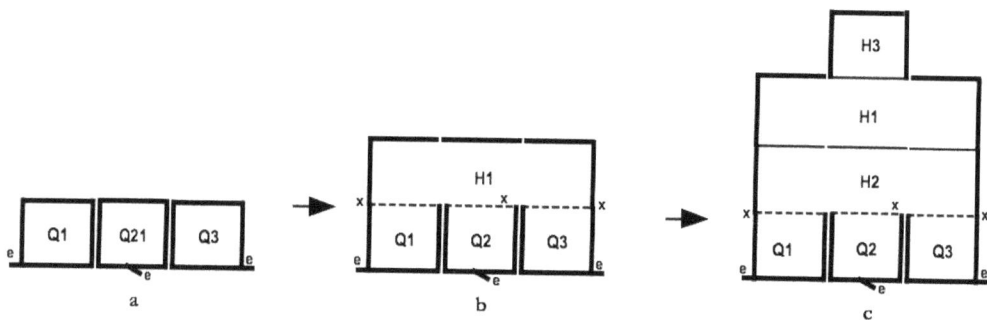

Figure 1.19 Coffin hive setup and operation: e = entrance, x = excluder, Q = queen; H = honey excluder:

(a) pairs of three strong single-queen ten-frame colonies were assembled each strapped together, six hives (two units) to the pallet;

(b) each triple queen unit was then united by papering on, above individual queen excluders, a thirty five frame full depth (coffin) super and covered with individual migratory lids to form a tripled hive; and

(c) each tripled hive was dynamically harvested, further supered with one or more coffin supers, even with the odd additional ten frame super (H3), during the long and heavy flows.

On arduous five day round trips, trucks ferried replacement gear to the many apiaries, returning filled supers to the tiny town of Wanaaring for extraction. The whole operation was designed to optimise the use of extracting equipment, supers and available labour and to harvest the maximum amount of honey.

Duff and Furgala[69] also refer a system of horizontal two-queen hives. Their hives were strong colonies split and placed on separate stands each supered separately. As such they were neither doubled nor two-queen hives. However the honey yields of their split hives were usefully compared to those of similarly treated un-split hives and to those of vertical Farrar-style two-queen hives.

An attempt to patent the general design of multiply queened seems, in retrospect, to be absurd, though details furnished in the patent provide some excellent insights into the actual construction of doubled hive setups[70].

For the regular bee owners with an interest in keeping bees for keeps' sake, one might be more content with keeping healthy gentle bees, maintaining good hive nutrition, providing adequate brood and honey storage space and managing bees to avoid swarming in regular single-queen hives. That way, and in a good year, you will always have enough honey for the table and that honey store.

However, like George Wells, keen to get the most honey possible from a few well kept hives rather than a sea of hives, the challenge is to spend more time with the bees and less on handling bees that struggle to maintain themselves. Canberra Region Beekeepers doubled queen setups (Figure 1.20) all adhere to the basic Nabors-Hesbach pattern of supering of pairs of single-queen hives at the earliest opportunity in late winter or in early spring. In practice they and two-queen hives (see Part II) have proved to be far and away the most productive hives in the club apiary.

a   b

Figure 1.20   Contemporary doubled hive variants:

(a) Canberra Region Beekeepers doubled hive employing eight frame and ideal full depth brood boxes supered with ten frame honey supers; and

(b) club member John Robinson's ten frame doubled hive with temperature and humidity sensing of brood nests to monitor nest development.

Photos: Alan Wade and John Robinson

Future improvements to the operation of doubled hives lie mainly in better integration of the modern doubled hive system with any of the myriad systems of autumn requeening and operation of doubled hives over winter. A member of Canberra Region Beekeepers, Dannielle Harden, is currently trialling a doubled hive setup (Figure 1.21) to more effectively overwinter her bees over Canberra's long winter.

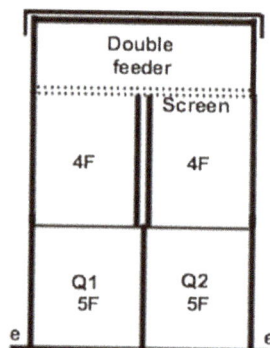

Figure 1.21 Overwintering nine-frame doubled hives (2021) with offset internal schematic: e = entrance; Q = queen

    Photo: Dannielle Harden

So does doubled hive beekeeping have a good future? It will certainly remain an attractive option to sideline beekeepers operating out of permanent apiary sites with early and good follow up honey flows and for commercial apiarists blessed with access to sites characterised by long and sustained honey flows. In most respects doubled or multiple single-queen hives can be managed in exactly the same way as one would manage any single-queen hive, the imperatives being to add more supers than one would normally do, to keep on top of removing and extracting honey and to take care to ensure queens are not accidentally transferred to neighbouring brood nests.

For temperate climes future improvements to the overall operation these hives lie mainly in better integration of the modern doubled hive system with any of the myriad systems of autumn requeening and operation of doubled hives over winter.

Part II – Two-Queen Hive Beekeeping

They form their combs and collect wax, an article that is useful for a thousand purposes of life; they are patient of fatigue, toil at their labours, form themselves into political communities, hold councils together in private, elect chiefs in common, and, a thing that is the most remarkable of all, have their own code of morals.

Natural History
Pliny the Elder[71]

The inventor of the two-queen hive, E.W. Alexander 1908

I – The Origin of the Single-Queen Hive

It is the first hatched queen that puts the others to death.

Natural History
Pliny the Elder[72]

In examining the essential nature of two-queen hives, we turn first to the origin of the single-queen hive. We then examine the exceptional circumstances under which a second queen may be maintained and efforts, dating back to the early 1900s, to introduce and maintain a second queen. In practice, the operation of two-queen colonies has been chastened by their readiness to return to the single-queen condition irrespective of their managed state.

However finding a transitory second queen is not uncommon as it routinely occurs when colonies are requeening themselves, that is when an old or deficient queen is being replaced directly. Further there may be many incipient queens in colonies preparing to reproduce in the process we call swarming.

The notion of running an extra laying queen to supercharge colonies has long lured apiarists into the belief that doing so would be straightforward. However the maintenance of a second queen has been so beset by uncertainty that the majority of beekeepers have abandoned all plans to do so.

Yet is any plan for running two-queen hives really so insurmountable, especially given we now know so much about their operation? The quite recent development of the consolidated brood nest two-queen hive, where two queens are present in the same brood nest but kept apart by a queen excluder, has simplified their operation before and during honey flows. If we add in the fact the normal swarming tendency of bees is strongly suppressed by the presence of a second queen – there are two lots of queen mandibular pheromone – one stills ponder why two-queen hives have not been more widely adopted.

Nevertheless harnessing the power of a second laying queen is routinely practiced. Nucleus colonies kept to back up failing queens and splitting hives to prevent swarming are but two examples of running extra queens in separate colonies intended to maintain the strength of single-queen production hives. Such supernumerary colonies can then be either united at the commencement of the honey flow or indeed at any time after disposing of one queen. This results in the combined forces of two laying queens being realised.

While queens of all honey bees in the *Apis* genus mate with many drones, likely more than twenty for *Apis mellifera*[73] and even more for some other species, the monogamous (single-queen) condition is almost always adhered to. So what are the origins of the highly evolved eusocial honey bee colony and how might we capitalise on the rare condition where a second laying queen is present.

Honey bees of the tribe Apini (*Apis* species) appeared in the early Oligocene[74] with the fossil record (Figure 2.1) spanning the Miocene (5-23 mya) and the Oligocene (23-34 mya). The similar morphology of early worker bees and modern honey bees strongly implies that ancestral colonies had much the same social structure.

Figure 2.1   Extinct *Apis henshawi* (Henshaw's Honey Bee) from the late Oligocene (~23 mya) in the Creské Stredhori Mountains of the Czech Republic[75]

The broader family (Apidae) includes the frequently eusocial corbiculate, so-called pollen basket, bees with a common progenitor dating back around eighty five million years (Figure 2.2). Queens in both honey bees and other tribes of pollen basket bees evolved to co-opt the support of daughters to raise and provision brood. A small number of less related hallictid bees have independently developed a less advanced sociality where young larvae are mass provisioned in a common brood chamber.

These traits maximised the potential of the queen to successfully raise new generations of bees carrying her genes, in the more socially advanced species entirely removing the risks associated with her having to forage and provision her offspring. In a remarkable process of specialisation amongst bombinid and honey bees, the queen became the sole mother of all members of the colony. Stingless bees (Meliponini) can maintain virgin queens and may have more than one laying queen – or queens kept in reserve – but they adopt the same principle of limiting the number of functional laying queens[76]. The queen's sterile female worker offspring, closely related to her, are relegated to the roles of foraging and of tending offspring although the workers themselves retain a capacity to exercise control over much of activity and behaviour of the colony. Such has been the degree of specialisation in these communities that subgroups, such as forager, swarm initiator and storage bees, may act independently of any queen signalling. Devolved decision making is one of the many hallmarks of advanced (eusocial) insects as well as amongst a few animal communities.

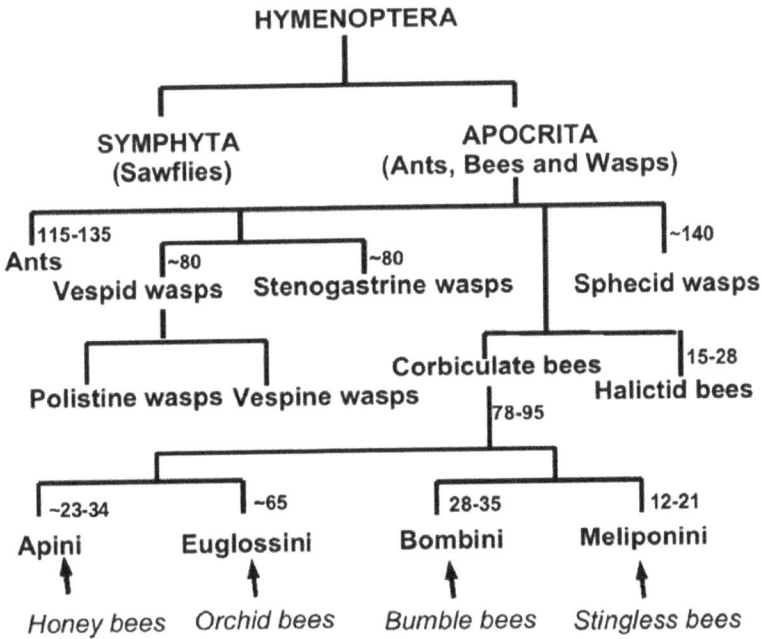

Figure 2.2   Simplified ancestral phylogeny of social ants, bees and wasps (millions of years ago)

In the process of social specialisation both Apini (honey bee) and Meliponini (stingless bee) colonies became potentially immortal though, in practice, wild as well as unmanaged colonies periodically fail. Only the individual bee is replaced making these insects a profoundly adaptable and efficient exploiters of floral resources. Further their propensity to accumulate stores has enabled them to radiate to more temperate climes and to survive extended periods of dearth.

In all, eusociality has evolved at least eight times amongst the hymenoptera, four times amongst extant bee tribes: the advanced eusocial Apini (honey bees), the Meliponini (stingless bees), the primitive eusocial Bombini (bumble bees) and the now solitary or weakly social Euglossini (orchid bees) all dating back to single ancestral primitively eusocial bee[77].

Honey and stingless bee tribes have evolved not only to a point where several generations of individuals coexist – that is where they work cooperatively to perpetuate the colony – but also to a condition where colony division (swarming) results in the formation of budded off colonies. Honey bees are more advanced in that they swarm to become immediately independent of the parent colony and in that the daughter colony of a primary swarm is headed by the parent colony queen. Within climatic limits and variability in availability of floral resources honey bees became capable of rapid and widespread dispersal, a situation limited amongst stingless bees where the daughter swarm only gradually becomes independent of the parent colony.

In the early ancestral condition, a single queen must have mated with a single drone, a strictly monogamous relationship. This condition was later relaxed so that the queen mated a number of times (polyandry), a trait particularly pronounced among honey bees (Table 2.1)[78].

Honey Bee Species	Average ♂ matings
Micapis (Dwarf Honey Bees)	
Apis andreniformis (Black Dwarf Honey Bee)	10.5
Apis florea (Red Dwarf Honey Bee)	7.9
Apis (Medium-sized Honey Bees)	
Apis mellifera (Western Honey Bee)	11.6*
Apis cerana (Asian Honey Bee)	14.1
Apis koschevnikovi (Koschevnikov's Honey Bee)	13.3
Apis nigrocincta (Philippine Honey Bee)	40.3
Apis nuluensis (Malaysian Mountain Honey Bee)	
Apis indica (Indian Plains Honey Bee)	
Megapis (Giant Honey Bees)	
Apis dorsata (Giant Honey Bee)	44.2
Apis binghami (Indonesian Giant Honey Bee)	
Apis breviligula (Phillipine Giant Honey Bee)	
Apis laboriosa (Himalayan Cliff Honey Bee)	16.8

*   More recent estimates suggest that thirteen drones may be needed to supply sufficient volume of sperm but that there may be many more matings.

Table 2.1   Honey bee polyandry

At first the notion of a colony of super sister worker bees working cooperatively with less related sisters to raise offspring, including future queens and drones only partly related to themselves, puzzled geneticists. However recognition that such colonies, comprising super sister cohorts, would be more likely to survive prevailed. This was explained by the fact that selective pressures are on the honey bee colony – as expressed by the queen – not on the individual worker bee. The strategy of queens mating multiple times – to maximise sperm storage – and to enable super sister cohorts to express special attributes, for example better colony defence, superior nest hygienic behaviour, better comb building capability and better foraging facility[79] all contributed to the survival of the colony and its queen.

The queen, totally focussed on egg laying and less prone to predation, averted the problem faced by other social insects such as vespid wasps and bumble bees. There the single mated queen is the founder of new colonies and must provision her offspring at annual startup with all the risks this presents to a fledgling colony.

Wild honey bee colonies are small in size when compared to those of contemporary commercial stock selected for honey gathering and pollination potential. For example, some African races of the Western Honey Bee (*Apis mellifera*) are poor comb builders and swarm prolifically, partly in response to pollen flows[80]. Indeed the reason that both the Western Honey Bee and the closely related Asian Honey Bee (*Apis cerana*) have been able to adapt to cold high latitude climates is that, in order to survive long winters, they both evolved to building their numbers quickly and to store surplus honey and pollen reserves. This aside, all honey bee colonies, for example the giant honey bees that migrate seasonally, must build quickly to enable them to swarm and establish new colonies.

Amongst the known dozen extant honey bees, pheromonal signalling guides the maintenance of the single-queen colony, a feature almost certainly shared with ancestral and now extinct species.

The first record of the propensity of queens to fight and to ensure that only a single queen remained was described in François Huber's 1790 Letter IX to M. de Réaumur[81] on the topic of the formation of swarms. Hugo von Buttell-Reepen[82] as well as Vivian Butz and Alfred Dietz[83] provide graphic and detailed accounts of the process of queen elimination and the circumstances under which it occurs. Hugo von Reepen famously reported:

If two queens come upon each other, only one will remain on the battlefield. If the queen is pleasing to every bee in every colony, the same thing may be said of the drones, who are extremely cosmopolitan, and who loaf about from hive to hive, and in consequence, apparently, of their specific odor, they are received peacefully everywhere, provided, of course, that the killing of drones has not yet begun...

Natural two-queen honey bee colonies

So we conclude that the normal condition for the honey bee is that of the colony being headed by a single queen only[84]. However queens, like workers and drones, all age and must be replaced periodically. So there are periods of changeover where many gynes (incipient queens) and drones (potential mates) are produced.

The droneright condition may persist all year but is generally restricted to conditions where queens can be raised or where nest conditions are particularly favourable. More typically colonies are perpetually queenright. However, and surprisingly, occasionally a multi-queen condition may be maintained if fleetingly. It is instructive to examine all the conditions under which queens are replaced to see how, and under what conditions, colonies maintain an additional queen.

The process of emergency queen replacement involves reprogramming a number of worker larvae, typically rather more than five in number not always very young larvae, to become queen bees. This occurs with any sudden loss of the colony queen. Unlike queen cells started from scratch on the comb surface or on the comb margin in other more orderly processes of queen replacement, emergency queen cells show their tell-tale worker cell origin in the worker comb matrix. Such emergency queens tend not to be of the best quality and may in turn be superseded.

One characterising feature of emergency queen replacement is that there is always an extended period, a period of about three weeks, during which the colony has no laying queen, that is where the colony is functionally queenless. Despite the setback the colony survives but only with a single replacement queen.

The process of supersedure, simple replacement of an ailing queen, involves the production of only 1-5 queen cells often, but not always, located near the centre of the brood nest. The surviving and newly mated daughter queen and mother, occasionally a total of three queens[85], may lay for an extended period, and is of not uncommon occurrence[86].

Now and then a second supersedure queen may result in a small swarm being cast with a virgin queen but this is an aberrant outcome. That both a founding and a daughter queen may coexist[87] is likely attributable to the fact that the old queen not only has low venom titre, and a low propensity to sting and hence defend herself[88], but also a weak pheromone signal, one that sometimes goes unrecognised by the young daughter queen and worker bees. More often than not however the young queen prevails and the colony is returned to its single queen condition.

One of the earliest records of the existence of sustained presence of two queens in a hive comes from Gilbert Doolittle's 1889 book *Scientific Queen-Rearing*[89], providing an account of queen supersedure:

> *After I had this experience with the colony that had two Queens in a hive (which was a surprise to so many fifteen years ago, when it was thought that no colony ever tolerated but one laying queen at a time), I began to watch for a like circumstance to occur, which happened about a year from that time. In the latter case, as soon as I found the cells, they were sealed over, and not knowing just when they would hatch [sic emerge], I at once cut them out and gave them to nuclei. In a few days I looked in the hive again, when I found more cells started, which were again cut off and given to nuclei, just before it was time for them to hatch. In this way I kept the bees from their desired object for some two months, or until I saw that the old Queen was not going to live much longer, when I left one of the cells, which they had under headway, to mature. By this plan I got about sixty as fine Queens as I ever reared, and laid the foundation for my present plan of securing Queens...*

Others, for example the notable Dr C.C. Miller, made observations on queens raised under induced supersedure conditions[90]. The simple procedure of isolating brood using queen excluders normally results in the establishment of a second or even a third colony queen provided entrances are provided to allow emerging queens to mate. This clear-visioned observation extended to the recognition that the process did not induce swarming as queens are raised under the queen pheromone deficiency supersedure impulse – not the overcrowding swarming impulse.

By 1921 reports of the presence of two queens in the one colony were so common that the editors of the *British Bee Journal* were prompted to exclaim[91]:

> *It is quite possible that there had been two queens in the hive, a condition that has been prevalent this year, or beekeepers are becoming more observant.*

With supersedure, there is little or no interruption in the brood cycle and most often the old queen is lost either just before or just after the new queen is established.

In several instances I have found that the old queen continued to lay well, an emerging daughter queen failing to replace her mother possibly signalled by vastly improved late spring honey flow conditions. Hepburn and Radloff[92] record instances where the two-queen status of colonies persists in African races of the Western Honey Bee including the Cape Honey Bee *Apis mellifera capensis*, the Arabian Honey Bee *Apis mellifera jemenitica* and in the African Honey Bee *Apis mellifera scutellata*[93]. Mathis[94] records uncharacteristic supersedure two-queen-colony persistence in the Tellian or Tunisian bee – also called Punics (*Apis mellifera intermissa*).

The process of swarming results in the production of a quite a number of queens[95] though the periodic formation of queen cell cups resulting from crowding is unrelated and arises from buildup of footprint pheromone[96]. *Apis mellifera lamarckii* (Lamarck's Honey Bee) is famous for producing between several dozen and several hundred queen cells[97]. In his travels Brother Adam[98] noted large numbers of swarm cells amongst the Syrian Honey Bee (*Apis mellifera syriaca*) as well as some strains of the Carniolan Bee (*Apis mellifera carnica*) and their excessive propensity to swarm.

Depending on the number of swarms issued, the parent colony will also experience a period where there is no laying queen or may indeed swarm itself to death. Nevertheless is not uncommon for there to be a number of nearly fully developed pre-emergent queens and even a number of free roaming virgin queens once a first or prime swarm departs.

Surfeit virgin queens remaining in the parent colony after all swarming has ceased are quickly eliminated so that, in line with the evolved condition, only one queen survives. After swarms issuing with a number of virgin queens may split and settle as independent new colonies each establishing as a single-queen colony.

Apart from the normally temporary arrangement of two-queen colonies arising from queen supersedure, there appear to be no scenarios amongst European honey bees where two-queen colonies arise spontaneously as a result of queen replacement. In a remarkable insight into the differences between swarming and supersedure Hogg noted[99]:

> *Queen supersedure is never the cause of swarming. But supersedure may occur concurrently with the swarm whenever the queen's failure was caused by being over-taxed while generating the bees for that swarm. Significantly, it is the*

*virgin queen that is then selected to accompany the swarm. The failed queen is
retained in the parent hive to be superseded in turn...*

*Apparently, just as the survival instinct of the bees in control won't allow a
swarm to leave a hive without a replacement queen in the parent [hive], they
won't allow a failing queen to issue with that swarm.*

While the propensity of any colony to default to a single queen state is always very
strong, it was such insight that eventually allowed John Hogg and, earlier, Eugene
Killion[100] to devise methods of crowding bees so that that were unlikely to swarm.
As we shall see Hogg was able to devise a scheme of running stable two-queen
colonies where the natural antagonism between queens was eliminated and where
the cooperative foraging behaviour amongst worker bees was optimised.

Parthenogenic requeening

In rare instances workers will spontaneously produce fertile offspring in a process
termed thelytokous[101] parthenogensis[102]. Such queenless colonies may and often do
requeen themselves. Even in this circumstance only a single queen survives.

Mackenson[103] and Butler[104] document a record of such instances with the Italian
Honey Bee, *Apis mellifera ligustica*, the Caucasian (from the Caucasus) Honey Bee
Apis mellifera caucasia, the Syrian Honey Bee, *Apis mellifera syriaca*, the Tunisian
Honey Bee, *Apis mellifera intermissa* and the Cape Honey Bee, *Apis mellifera
capensis* from South Africa. The trait is so pronounced in the Cape Honey Bee that,
according to Butler, colonies that become queenless are reluctant to raise emergency
queens. Such colonies continue to function normally, laying workers – pseudo
queens – producing both normal workers and small drones raised in regular worker
comb. However under nearly all circumstances greater than 99% of European races
of colonies of the Western Honey Bee will dwindle and perish as will many Cape
Honey Bee colonies. Interestingly drones from hopelessly queenless colonies retain
the potential to pass on the former queen's genes, that is assuming they can then
mate successfully.

Smith (1961)[105] makes the earliest reference to the existence of fertile laying workers
in the Tellian Honey Bee (*Apis mellifera intermissa*):

*Towards the end of this period, bees from Tunis, which were referred to as
Punics, caused some interest in Britain, where they were first imported by
J. Hewitt, who wrote under the name A Hallamshire Beekeeper. John Hewitt
(1892) reported that Punic bees could rear queens from the eggs of laying*

53

workers, and this appeared to be quite a common occurrence with his strains.

Apart from many confused reports disputing the existence, temperament and productivity of these Punic bees (*Apis mellifera intermissa*) outside their native Tunisia and in the near Middle East, Hewitt[106] made findings of these parthenogenic bees almost a decade earlier though Hewitt's claims were later roundly dismissed[107]. This was not a case of workers laying and having queens raised parthenogenetically but rather workers laying multiple eggs in a cell alongside fertile queens from which drones were raised in worker cells. It appears that this observation was wrongly ascribed the queen to being a drone layer. The well-respected Baldensperger[108] gave no credence to the Punic or Tunisian Honey Bee (*Apis mellifera intermissa*) or the Syrian Honey Bee (*Apis mellifera syriaca*) being able to requeen parthenogenetically.

The more celebrated discovery of parthenogenesis amongst Cape Honey Bees (*Apis mellifera capensis*) dates mainly from the detailed observations of Onions[109]. His findings, made between 1912 and 1914, attracted wide commentary[110], but it was not until the 1950s that any systematic reviews[111] of Cape Honey Bee biology were undertaken.

It is very clear from the 1857 writings of the eminent physiologist Carl Theodor Ernst von Siebold[112] known well to Johann Dzierżoń[113] (the Father of Beekeeping) that such fertile unmated queens were entirely unknown to the beekeeping literati, that is prior to the discovery of the aberrant self-queening behaviour in the Cape Honey Bee.

Recombinant primary swarms, those merging from different colonies, each containing an old laying queen occasionally settle to form a new colony headed by two queens. This condition appears rare and such swarm colonies eventually revert to the single-queen colony status[114].

Parasitising two-queen colonies. The existence of two-queen colonies and the aforementioned self-requeening of queenless colonies is pronounced in some races of African strains of the Western Honey Bee, particularly in the aforementioned Cape Honey Bee, *Apis mellifera capensis*[115]. Single queen Cape Honey Bee colonies bees also drift to or swarm into African Honey Bee (*Apis mellifera scutellata*) colonies sometimes resulting in the formation of two-queen colonies[116]. There the African Honey Bee queen is relegated to the nest periphery and the colony is overtaken by Cape Honey Bee laying workers. Such colonies, and indeed whole African Honey Bee apiaries, have perished[117] leading to controls being placed on Cape Honey Bee migratory beekeeping.

Magnum[118] provides some evidence that late swarms of European honey bees supplant existing honey bee colonies or progressively rob them thus parasitising their resources. However this kleptoparasitism does not lead to the formation of two-queen-colonies. This phenomenon has been most widely reported on for takeover of the African Honey Bee (*Apis mellifera scutellata*) by the Cape Honey Bee (*Apis mellifera capensis*)[119].

The reproductive biology of these southern African races of honey bee is complex but the fact that two-queen colonies commonly exist confirms that multiple-queen colonies may, a priori, be operated successfully even where queen excluders are not used.

As already noted, amongst normal managed colonies employing queen excluders, the appearance of a second laying queen is common where brood nests are split and where a second hive entrance is provided allowing an additional queen to be raised and to mate[120]. This sometimes occurs in practicing the Demaree swarm control plan[121] where brood, separated from the queen, raises queens one of which may mate and establish a separate brood nest.

The queen breeder Jay Smith pointed out such reversion most often occurs sooner rather than later[122]:

> *I learned it was not difficult to introduce queens to each other, so they would be friends, yes, regular old cronies, always working together, and usually found on the same comb. The only discovery of importance was the fact that it is the bees that make plural queens in a hive impossible.*

What was not known was the way multiple-queen colonies needed to be managed to prevent their collapsing to a single-queen colony, that is preventing one queen superseding all others. It came as a surprise to many when it was discovered that two-queen colonies could be established and maintained. That some skill and a large effort would be required to help them reach their full potential emerged as the main early barrier to their adoption.

II – The First Two-Queen Hives

As he listened he heard more than once the rustle and slide of a honey-loaded comb turning over or falling away somewhere in the dark galleries; then a booming of angry wings, and the sullen drip, drip, drip, of the wasted honey, guttering along till it lipped over some ledge in the open air and sluggishly trickled down on the twigs.

The Second Jungle Book, Chapter 13 – Red Dog

Rudyard Kipling 1895

The first American two-queen hives

In a letter to the *Gleanings in Bee Culture* published on the 1ˢᵗ of April 1907 E.W. Alexander[123] provided an outline of a scheme for successfully introducing a number of queens, as many as fourteen, to a colony. So startled was the editor that he prefaced Alexander's letter with a cautionary note:

> *If a beginner or some old bee-keeper unknown were to write along the lines in the subjoined article the average apicultural editor would have been inclined to turn it down and say that such writer had, perhaps, better get a little more experience before wading out so far into deep water. But Mr Alexander is no novice in the honey business. His annual crops go up into carloads, and his experience covers a lifetime. He has given a great many valuable hints. We hope, therefore, that the reader will not be ready to exclaim: 'Impossible! Absurd! Nonsense!'...*

To add to the drama Alexander noted that his son had also developed an entirely new method for sucessfully introducing such additional queens:

> *Last summer my son Frank discovered the most practical method of introducing queens that I have ever heard of – a method whereby over 90 per cent are safely introduced and laying within 18 hours from the time the parent queen was removed...*

The editor of the *Gleanings in Bee Culture*, E.R. Root, continued to support Alexander. As a prelude to Alexander's followup bomb-shell letter of September 1907[124] this same editor commented:

This is the long-expected article for which many of our readers have be anxiously waiting. Now that it has come, some of the statements are so startling that, had they come from any less authority than our correspondent, we should feel inclined to lay it aside to think it over, if not pigeonhole it altogether. But instead of saying 'No it won't work', our readers are requested to try it and report. The immense possibilities that might accrue from the use of two or more queens in the one brood-nest are too great to be lightly dismissed.

Alexander furnished further details of how he established multi-queen hives. To achieve this seemingly impossible feat he employed an elaborate engorgement and preconditioning technique to introduce an extra queen (or queens) to an already queenright colony. Alexander[125] was soon to provide a postscript to his startling revelation to which the editor further helpfully added:

... 'For', we said, 'we would suppose that, after the honey-flow had stopped, and there was a strong disposition on the part of the bees to rob, one or more of the queens would disappear until only one was left'. We now raise the question whether it is practicable to practice this dual or plural queen system, even with the use of perforated zinc, after prosperity has begun to wane. It is then, according to our experience, that there seems to be a strong disposition on the part of the colony, or the queens themselves, to have only one mother in the hive. Mr Alexander's theory, that when the queens have stopped egg-laying, and therefore they are more active, may explain why the queens are belligerent enough to fight to a finish. But we are of the opinion that the bees themselves take a hand in the matter, on the principle that economy and retrenchment are the order of the day.

Clarence Hall[126] was soon to experiment with and report on Alexander's technique to which the editor of *Gleanings* invited a response from Alexander. Alexander graciously provided details of the technique for queen introduction that his son had discovered footnoting Hall's letter with the detail:

The main and most vital part of this method seems to be in introducing the strange queen to about a quart of bees that are well filled with honey after they have been taken from their parent colony a few hours, and keeping them for a few hours longer with the queen you wish to give to the colony. This gives the queen the odor of the colony, and the bees don't seem to realise but that it is their mother queen. All the rest is but secondary to this main part. E.W. Alexander, Delanon, N.Y., Oct. 14.

Here it is worth drawing attention to the erroneous notion – still held amongst beekeepers today – that hive odour is a factor in determining queen acceptance[127]. However it is clear that Alexander's queens were in such a replete queen condition that they would be in an ideal condition for acceptance. The contemporary practice of feeding bees to improve acceptance of queen bees in normal requeening operations – especially in the absence of a light flow – is testament to the value of this practice.

Alexander[128] concluded that:

> ...We can now safely introduce any number of queens to a colony that has a laying queen and is in a normal condition.

In reviewing Alexander's *plurality of queens*, Macdonald and Avery[129] drew much the same conclusions, Macdonald noting that:

> It is only quite recently that any bee-keeper was known to assert that two or more queens can exist in one hive, laying peacefully side by side. The subject is an exceedingly interesting one, and if successfully proved to be feasible it may revolutionise beekeeping. Gleanings has two articles on the subject. Mr Alexander – a high authority – positively asserts that he found five queens in one hive quietly performing their functions without any feeling of rivalry. Further, he gives a method by which he asserts several queens may successfully be given to any hive, and you will not lose one queen in a hundred. Mr J.E. Chambers adopts the opposite view, and maintains that although he has been using more than one queen in a hive to build up, it must be under proper safeguards – something, I take it, like our Wells dummy to divide the two domains. Mr Alexander's claims he sets aside as impracticable, if not impossible.

Despite Alexander having yet to fully divulge details of his method of running two-queen or multiple-queen hives, a meeting of The National Bee Keepers' Convention[130] held in Harrisburg in late October 1907 put a seal on Alexander's discovery stating:

> ...that there seemed to be a general agreement that two or more queens, each one separated by perforated zinc from every other queen, could be kept in one colony of bees so long as there was general prosperity in the hive; but when a dearth of honey came on, there seemed to be a feeling that all the queens would disappear except one. Two or three reported they made a success of the two-queen system. Among them was Mr E.E. Pressler, of Williamsport, Pa. He, like Mr Alexander, had made a success of it, and even gone so far as to make the system work without the use of even perforated zinc, but had not

been able to test the principle this past summer, owing to an affliction of the eyesight; but he thought there was great possibility along that line. But the majority of those present who took part in the discussion seemed to feel that it was practicable to run two queens to a hive, providing they were separated by perforated zinc.

This reference to E.E. Pressler is elaborated on in some detail under Editorial in the May 1 and November 15, 1907 issues of *Gleanings*[131]. The editors record that Pressler, and later Green, had replicated Alexander's discovery in that they had been able to operate two-queen colonies excluder free. These reports were corroborated with another report from Alexander[132] of five queens being maintained in a hive into winter where a further editorial response also recorded the successes of these apiarists.

It is hard to reconcile the doubts expressed by beekeepers of the time about the feasibility of operating multiple-queen hives, especially those operated without queen excluders. In a final March 1908 communication to *Gleanings*[133] published in September 1908, Alexander lamented the regressive attitude of many to his system though he pointedly promoted a requeening scheme that is today in common practice:

> *All that is necessary is to form our nucleus over a queen-excluder on any colony of medium strength, and the next day introduce a laying queen to the nucleus; then in about 25 days the hives can be separated. Each hive will be full of bees and maturing brood; then move the lower hive a little to one side, and set the upper hive alongside so about an equal number of the working force will enter each hive. Here you now have two good strong colonies in the place of one. Each has its laying queen, its hive filled with brood, and a good working force of bees.*

In a telling footnote to a 1911 letter by A.B. Marchant[134] the editors of *Gleanings* attributed much of the success of Alexander's and Doolittle's pursuits – including those of operating hives with additional queens – to their consummate beekeeping skills and to the extraordinary floral resources of their beekeeping districts. Alexander is credited with harvesting over thirty seven tonnes of honey from 750 hives (Figure 2.3) located at a single site so could rail against the prevailing belief that multi-queen colonies could not be maintained.

THE ALEXANDER APIARY AT DELANSON, NEW YORK. ONE OF THE LARGEST APIARIES IN THE WORLD

Figure 2.3   E.W. Alexander's apiary[135] run by his son Frank 1921

Others, such as Chambers, Hand, Wright, Davenes, Sherrod, Bussy, Gray, Joice and Simmins[136] were soon to follow in successful operation of two-queen hives though others such as Robinson and Whitney[137] decried their use.

In a recent two-part series[138] I compared the performance of doubled and two-queen hives noting, in rather more detail, Alexander's success in establishing any number of queens in a hive without a queen excluder.

The first United Kingdom two-queen hives

In a letter dated the 5th of August 1907 – but not published until December of that year and prior to Alexander's September release of his detailed plans – Cruadh[139] indicated that he had a well established approach to two-queen hive beekeeping in Ireland:

Ever since Mr Alexander's article [Gleanings 1907, p.473], I have watched your columns carefully with the expectation of seeing his promised subsequent article. Others, too, at your request, have given their experiences, but, like Mr Alexander, all have refrained from detailing their methods. The writer far from wishes to tread on the tail of Mr A.'s coat, as they say in Ireland; but as the season is advanced I am to-day giving a method — the best of the many possible ways by which this object may be attained, so that it may be tested at the Home of the Honey-bees.

We are daily using this Cruadh (so called) method of introducing, which is an infallible means of introducing any queen to any colony, and that, too, in a way which not only embodies all Mr Alexander claims, but more also, for the introductions, even in plurality, never fail. There is no necessity for removing the old queen; the work of the colony is in no way impeded, and the new queen or queens will be laying in just about the 18 hours mentioned...

So while it would appear that operational two-queen hive systems were well established by late 1907 in both America and the United Kingdom, it is quite certain that Alexander was the inventor of the scheme to introduce queens to excluder-free queenright colonies. In hindsight Alexander's finding fits in well with the much later discovery that two-queen colonies often continue to operate with both queens for a period after all excluders are removed. The jury is out as to who first operated two-queen hives though it seems likely that they were invented independently in America and the British Isles.

By 1908 both Medicus[140] and Ellis[141] had successfully adapted Cruadh's two-queen management scheme to the late heather honey flow. Since strong colonies are essential to maximise a honey crops and since heather blooms when bees are in seasonal decline, any two queen setup would have provided a ready solution.

Using two queens, they hauled in very large crops of the very valuable late-flowering Ling Heather (*Calluna vulgaris*) honey. Not surprisingly, we learn that the bees from the Landes region of southwestern France had themselves long worked out how to exploit heather honey but with a single queen. There the local strain of the dark European Honey Bee *Apis mellifera mellifera* has an atypical second end-of-summer brood rearing peak[142]. The bees, not just the beekeepers, had discovered how to harvest heather blossom honey efficiently.

In a forward to his more fulsome 1910 account Medicus[143] made an important observation on the Wells' doubled hive scheme distinguishing it from the two-system where he noted:

...there has nearly always been misunderstanding as to what is meant by a two-queen system, owing to its confusion with the system advocated by the late Mr George Wells... Mr Wells did not advocate working a colony with two (or more) queens, but advocated giving to two stocks a super, or supers, to which both colonies had access. He [sic They] profited by the mutual warmth which two colonies in such close proximity must derive from each other, and this enabled them to build up more rapidly in the spring. This advantage was increased during the honey-flow, when, by having a super common to the two colonies, a larger proportion of nectar-gatherers were able to be liberated from home duties.

Medicus then went on to define what constituted a two-queen hive:

...A true double [two-queen] or multiple queen system is one in which a single colony with a single entrance[144] has two or more queens laying at the same time, and the workers of which have access to every part of the hive. The queens may be either loose in a common brood chamber, or kept apart from each other by queen excluders.

He attributed his setting up two-queen hives to the aforementioned Irish apiarist Cruadh describing the technique for establishment of a second queen:

The simplest and safest method of introducing a second queen to a colony was pointed out to me by Cruadh, whose name used to be known to bee-keepers, to whom I am greatly indebted. Having a spare queen, go in the morning to the colony to which this second queen is to be introduced, and remove from it two or three combs of sealed and hatching brood, and place them in a spare brood-chamber division, in the front of which a ½-in. hole has previously been bored. Cage the new queen on one of the combs, and shake in enough bees to care for the brood, but care must be taken that the original queen is left in the original brood-chamber. Having filled up the empty spaces in the original colony with foundation or drawn combs, place over it a frame of wire gauze, and on this stand the nucleus just formed with its new queen. If this manipulation is performed in flying weather the new queen can be liberated with safety on the following morning, as all the older bees will have escaped by the upper entrance and have returned to their old queen below, and only young queenless bees will be built. The combs, however, should be separated on opening the hive, and a few minutes allowed for any of the older bees still left to take flight. Within twelve hours the newly-liberated queens will generally have begun laying.

We are fortunate in having not only his lucid instructions on the operation of two-queen hives but also to have Medicus original drawings for their setup (Figure 2.4) and for their deployment to the heather (Figure 2.5).

Colonies were first built in a process of regular brood chamber reversal until, well into the season, a second queen was introduced by establishing a nucleus colony split off from the hive and placed on top at the top above a wire screen. The colonies were then united by simply replacing the screen with an excluder, the two queens working together in a top-bottom brood nest arrangement to form exceptionally powerful colonies.

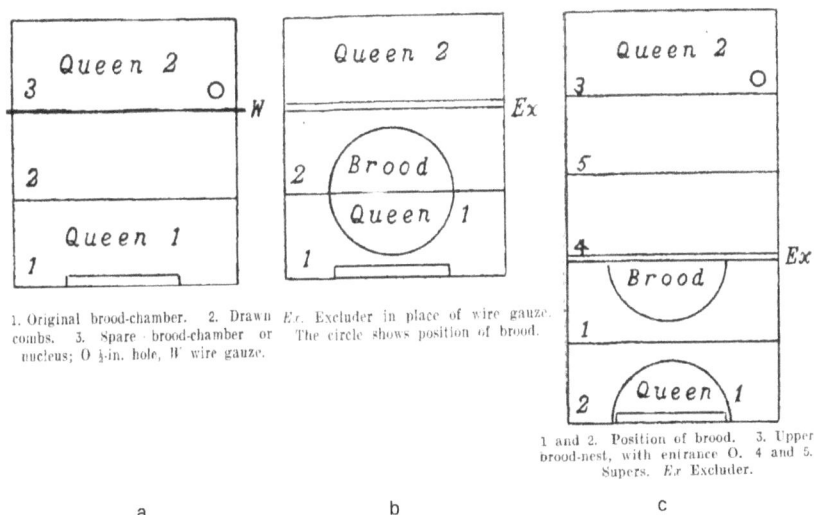

1. Original brood-chamber. 2. Drawn combs. 3. Spare brood-chamber or nucleus; O ⅜-in. hole, W wire gauze.

Ex. Excluder in place of wire gauze. The circle shows position of brood.

1 and 2. Position of brood. 3. Upper brood-nest, with entrance O. 4 and 5. Supers. Ex Excluder.

a   b   c

Figure 2.4   Medicus two-queen hive set up:
(a) a second queen is established above a dividing screen;
(b) the colony is united and consolidated by replacing the separating wire screen with a queen excluder; and
(c) brood chambers are reversed to accelerate buildup.

Just prior to the heather flow, the colonies were reorganised and split. The old queen was left in a small colony in the home apiary while most of the bees with sealed and emerging brood, and the new queen, were transported to the moors.

At the end of the flow, and with the heather crop removed, the now nearly spent bees were returned and reunited with their parent hives in the home apiary making them ready for winter.

Despite the vicissitudes of Scottish weather, these honey hunters obtained both good clover and heather honey crops. That these enterprising beekeepers had turned to two laying queens made this feat possible.

Queenless
brood

Ex

Queen 2

Ex

Queen 1

TWO-DECKER COLONY.
Ex, Excluders.

a

Queen 2
& young brood

HOME COLONY.

b

Sealed brood

Super

Super

Queen 1
& old brood

Ex

COLONY AT THE MOORS.
Ex, Excluder. The arrow shows position of hole in calico.

c

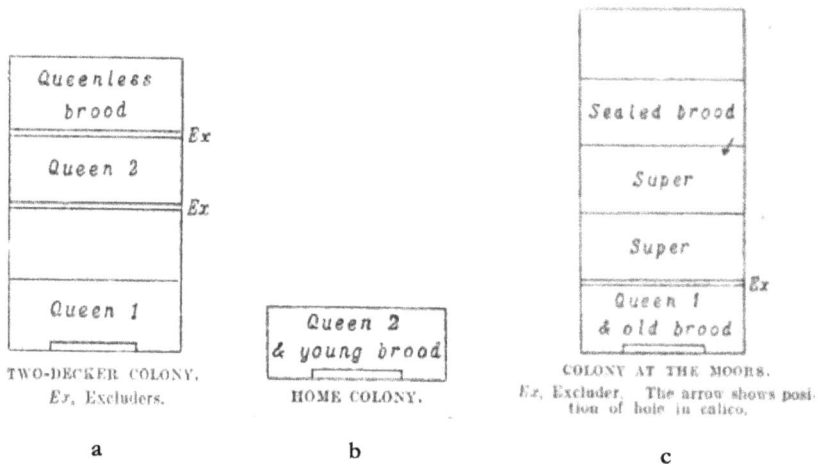

Figure 2.5 Medicus scheme for deploying most of bees and a single queen to the heather flow:
(a) the two-queen colony is reorganised just before the split;
(b) one queen with all the eggs and young brood is offset; and
(c) the colony with all the sealed brood and most of the bees and the remaining queen is moved onto the moors as the heather flow commences.

Ellis's scheme differed only slightly from that of his early collaborator Medicus. In this, Ellis made splits of strong colonies, starting again as Medicus had done, by piggybacking a nucleus with a new queen but using a split board instead of a screen:

Suppose you have a colony occupying two sets of shallow frames. Just as the clover flow begins contract to a single body of ten full frames of brood, supering with racks of sections. Above all place a large board, and on this the box of surplus frames containing small patches of brood with honey and pollen, at the same time providing a small entrance through the hive lift. By the following evening only a few young bees will be left, when a fertile queen can be safely introduced. We have the original stock now in two portions, one making the best of present opportunities, the other building up for the heather campaign.

This scheme was in part made possible by the lateness of the heather honey flow as there was ample time to build bees and make the two-queen venture economically feasible.

But these were early forays and it was not until the 1930s that Farrar, and later Moeller and Hogg, developed more ingenious methods for running two-queen hives, robust schemes with more general application. The American research effort, supported by the United States Department of Agriculture, was far more sustained than that of the southern Scottish venture and laid the basis for all future developments of two-queen hive systems.

The take home message from all early experimentation was that two-queen colonies may persist in that state, even after removal of all queen excluders, that was at least until autumn or until the main honey flow was over. Very much later and around 1950, Spoja[145] in Yugoslavia, corresponding with Kovtun[146] in the Kharkov area of the Ukraine in the then USSR independently explored the conditions under which multiply-queened hives could be established and maintained in the absence of queen excluders. Their findings, referenced by Farrar[147], confirmed Alexander's notion of hives hives operating with a plurality-of-queens.

By the 1920s it became apparent that the conditions of finding two queens in the one honey bee colony were contingent upon either:

- colonies undergoing queen supersedure where the hive tolerated mother and daughter occasionally for a sometimes extended period of time; or

- hive manipulation involving separation of brood leading to the presence of a second laying queen. Both daughter and parent queen may each lay in discrete brood nests; or

- a more determined effort to establish a second queen – usually a laying queen – as demonstrated by Alexander, Cruadh and Kovtun – where her introduction and establishment were carefully finessed.

Despite the overwhelming evidence that two-queen hives are far more productive than equivalent pairs of single-queen hives, in a poor season this is not always the case. Lensky and Golan[148] report that in drought years only marginally more honey is produced than in a single-queen colony. Importantly they also report that:

...each one of the two queens in a two-queen colony oviposited less than one queen in a single-queen hive.

The general condition of multi-queen colonies have been investigated by Lensky and Darchen[149], though the idea of clipping queen stings to maintain a second queen goes well back to the aftermath of Alexander's findings where masking techniques – e.g. smoking bees – were employed to introduce additional queens[150].

More recently multiple-queen colonies, those operated without a queen excluder, have been run year round albeit after disabling queen defences. For example, Paleolog and coworkers[151] also clipped queen stings and Zheng and fellow collaborators[152] ablated, that is blunted, the mandibles of queens. Using these queens, Zheng's group were able to maintain three-to-six-queen colonies over winter. Of more interest has been the finding that multiple queens can be maintained in colonies by masking chemicals daubed onto mandibles and antennae[153].

Woyke[154], Szabo, Wyborn and many others[155] have studied the successes and problems of queen banking, notionally multiple-queen colonies, including keeping mated queens over winter. However, queens under these conditions are always confined and, while in a potential laying condition, are functionally unable to do so. Other studies such as those conducted by Dietz[156] have focussed on problems of year round maintenance of two-queen colonies.

Expansive studies on the mating behaviour of queens, both within and introduced to hives, made by Gary[157] and Szabo[158] provide some clues to the behaviour of colonies with more than one queen and signal some of the problems attendant to maintenance of two-queen hives. For example Farrar[159] notes:

> *Swarming is less of a problem in two-queen colonies than in strong single-queen colonies, but queen cells started because of a failing queen or crowding in either brood nest will stimulate production of queen cells in the other.*

Healthy honey bees with optimal nutrition and a young queen have enormous honey gathering potential and are the backbone of commercial beekeeping[160]. The presence of a second laying queen or, as already covered in Part I, the parallel operation of two single queens in a strong hive split well in advance of a honey flow has the potential to further enhance colony production.

It is now time to trace the major developments in the operation of two-queen honey bee colonies commencing with the findings of Farrar. Despite all schemes employing two queens demonstrating greatly enhanced honey production, every scheme presented logistic challenges that have only been overcome recently.

III – Managed Two-Queen Hives

We bring it to the hive and, like the bees,
Are murdered for our pains.

Henry IV
Shakespeare

The 1930s and 1940s saw the emergence of a long line of researchers intent on developing better ways to establish and operate two-queen hives. The most eminent of these were Clayton Leon Farrar, Charles Harold Gilbert, Winston Edson Dunham, John Edward Eckert and Frank Robert Shaw.

Their efforts in handling overly large colonies with a complex mixture of brood and honey supers were heroic. While they were – like the pioneers – fully cognisant of the requirements to run such colonies, the skills to do so remained well beyond the capacity of the majority of beekeepers. It would however appear that the Farrar brothers really enjoyed engaging with their bees (Figure 2.6)[161].

Figure 2.6   Clayton and Milton Farrar, both professors of entomology, with other interests

In an interesting aside one of Professor Farrar's disciples, Dr Don Peer, had given up a very promising research career to become an extremely successful beekeeper. Eva Crane[162], the founder of the International Bee Research Association, had met up with Don at one of his apiaries in Canada and they were discussing the fact that his two-queen colonies were putting on a regular18 kg (40 lb) of honey a day. Eva wrote:

> *I saw his outfit and stood on the back of a truck to reach the top supers. Such tall hives made him switch back to single-queen hives, but even then he stacked supers as high as he could reach. 'Bees need space' he said.*

In a remarkable series of experiments Farrar was able to not only quantify the impact of colony strength on honey production (Figure 2.7a) but also explain how colonies build and why extra strong colonies are far more productive per bee than average strength colonies while weak colonies struggle to store any surplus honey (Figure 2.7b). He then went on to plot the trajectories of colony buildup starting with broodless bees (a swarm or package bees), an overwintered single-queen colony, and a colony split to form two colonies that were united as a two-queen colony (Figure 2.7c). How to build bees so that the population peaks coming into the main honey flow requires a very nuanced appreciation of the buildup process[163].

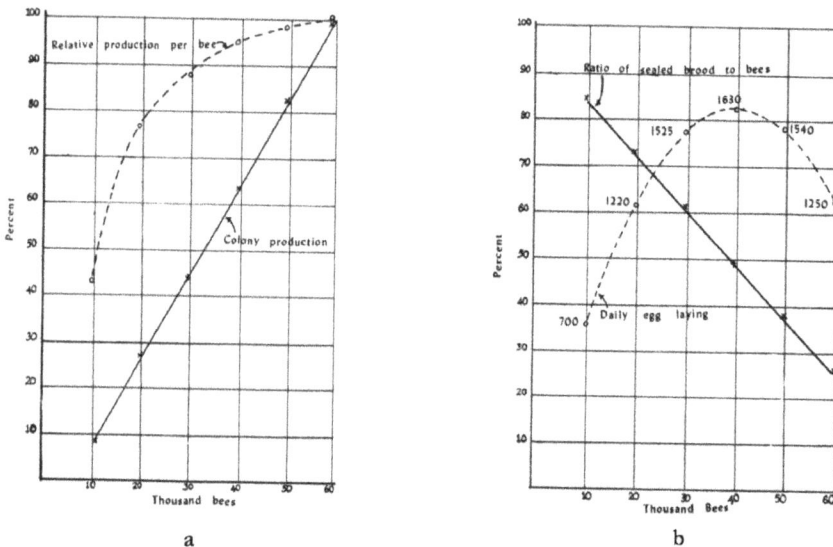

a

b

Figure 2.7   Farrar's plots of colony performance:

(a) colony production in relation to colony strength and increased productivity per bee as the population increases; and

(b) increasing ratio of total bee numbers to amount of brood (expressed as ratio of decreasing brood to adult bees) as the population expands freeing up bees to become foragers (linear relationship) and, for a young laying queen, daily egg laying rate increasing almost linearly until the population exceeds about 30,000 bees (parabolic relationship).

Farrar[164] put it this way:

In single-queen colonies the production per unit number of bees increases as the number in the colony increases up to the maximum (60,000 bees). This efficiency relationship remains high when populations are further increased through the use of a second queen.

Farrar had demonstrated the population response and nectar gathering potential of hives under different systems of management. He showed that the amount of honey gathered by bees could be simply linked to colony bee numbers[165] and so discovered why such two-queen colonies, well managed, greatly out-performed equivalent pairs of single-queen hives.

In his defining 1968 paper on maximising colony strength, Farrar outlined the factors controlling colony buildup:

The time factors expressed in numbers of days required for the three classes of colonies to reach a maximum strength and enhance their maximum production efficiency are shown in Figure 2.7c. These colony growth curves are based upon daily egg-laying rates, time of brood development, and the length of life for adult bees when healthy colonies are headed by good queens and abundantly supplied with honey, pollen, and hive space in a favorable position. The production efficiency per unit number of bees in the two-queen colony is equal to or slightly greater than that of full-strength single-queen colonies. When a measurable nectar flow develops about the time the second queen is introduced, the storing efficiency of the colony will be lowered, since more bees will be engaged in raising brood from the two queens. On the other hand, its production efficiency will be higher than that of single-queen colonies when united back to single-queen status. For 20 days such colonies have essentially double populations with brood from eggs laid by only one queen.

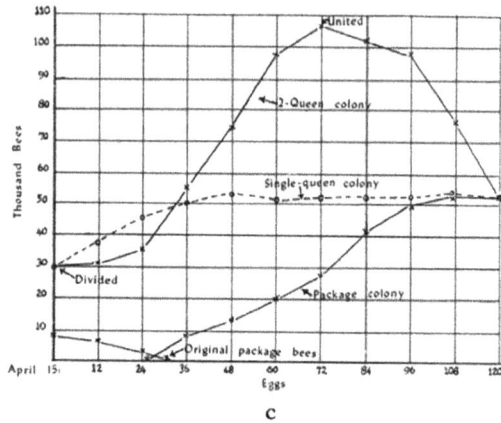

Figure 2.7 Farrar's plots of colony performance (continued):
(c) population expansion over two-three months under different systems of queen management again assuming highly fecund young queens.

Further factors that may boost bee numbers include using stock selected for queen laying potential and hygienic behaviour[166], optimising colony nutrition, hive insulation and siting of hives in sunny protected positions to minimise colony stress. Farrar had thus pioneered practical means of operating two-queens knowing that, with the services of an extra queen, colonies with a formidable honey gathering capability would be produced.

He further elaborated on the size of colonies needed to realise honey flows:

> *Good nectar flows are often not recognised because colony populations are too small to show gains. The producer of package bees can profitably manage colonies within the range of 10,000 to 20,000 bees, periodically shaking marked bees. This is because colonies with 10,000 bees raise proportionately more brood than larger colonies. The beekeeper who keeps colonies for honey production or plant pollination must direct his management towards developing all colonies to full strength for particular crops. Package bees used to establish new colonies require 11 to 13 weeks to reach full production efficiency. The division of strong colonies six to eight weeks in advance of the principal nectar flow to establish two-queen colonies is a means of obtaining relatively higher brood production early in the season and high production efficiency when nectar is available.*

In a revealing 1952 interview with the Editor of the *American Bee Journal* Bud Cale[167], Clayton Farrar distilled his findings on the operation of two-queen hives:

Colony populations must be built before the flow if a crop is produced, not on the honeyflow. Two-queen colonies take advantage of the principle that the production per unit number of bees increases as the population increases. Thus, they exhibit not only greater colony gains but also greater gains per bee that is reared.

One of the most-widely acknowledged early reports of successful operation of two-queen hives came with the publication of Dugat's book[168] *La Ruche Gratte: Ciel a Plusieurs Reines* – The skyscraper hive with several queens. Eva Crane[169] reports that the publication caused quite a stir when it was published in 1946 (an English translation[170] *The Skyscraper Hive* appeared in 1948). Butler, however refuted the publicity given to Dugat in stating[171]:

...a good many beekeepers will heartily disagree with M. Dugat's system of colony management. His system is indeed little more than an extension of the two-queen system of management which has been tried out very thoroughly in the United States and other countries since at least the beginning of this century, and has not been found to yield any notable increase in honey crop per queen over various single-queen systems.

By this time, however, Farrar[172] had not only already demonstrated that two-queen colonies could be readily set up and greatly increase honey yield, but had also developed the main practical means for their successful operation. His experiment spanned an eleven year period (1934-1946), a monumental study. Farrar was joined by John Edward Eckert (1937) and also by Charles Harold Gilbert (1938), later by Floyd Moeller (1961)[173], in realising the superior honey performance of populous two-queen colonies, Farrar being undoubtably the great pioneer of two-queen hives.

Farrar's experimental setup

Farrar's system employed shallow ten-frame supers, and in a parallel setup square shallow twelve-frame supers throughout, to maximise flexibility for rotation of supers. He emphasised the importance of establishing hives on firm level bases and his 1958 paper provides detailed year-round operational details required for their setup and management.

To set up his two queen colonies (Figure 2.8), Farrar first split strong colonies 6-7 weeks in advance of the main honey flow. He did this by moving up about half the brood, mainly sealed, above an inner cover with a separate top entrance and introduced a new laying queen.

Farrar summarised his approach in stating:

Two-queen colony management represents an intensive system of honey production designed to obtain the maximum yield from each hive unit during any two-week honey flow, the honey product on per unit number of bees increases with the colony. In other words, one large colony will produce more honey than two or more smaller colonies having the same aggregate number of bees. Since small colonies may increase their populations, under a long honey flow the relative production between the two classes of colonies need not remain static. Colonies with large populations throughout a honey flow, whether long or short, make the greatest total gain.

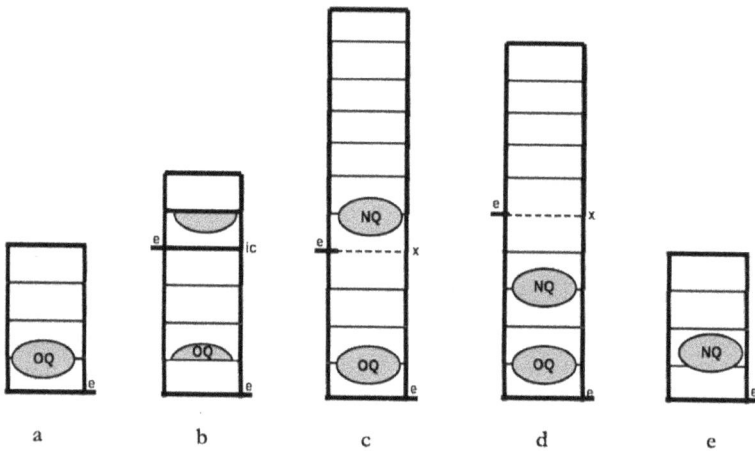

Figure 2.8 Clayton Farrar's 1946 system redrawn showing key steps in operation of two-queen colonies depicting twelve-frame shallow super operation: e = entrance; ic = inner cover with small screen; x = excluder; OQ = old queen; NQ = new queen:

(a) a strong single-queen colony employing a good previous season queen is built early in the season with brood supers regularly reversed to arrest swarming and to encourage the queen to lay at near capacity;

(b) the colony is split, a new queen is established in an upper unit above an inner cover and the lower unit is supered to facilitate expansion;

(c) the two-queen unit with new queen and old queen is further supered and dynamically operated during the honey flow;

(d) the brood and super chambers are rearranged towards end of the honey flow; and

(e) the colony is consolidated to a strong overwintering unit with all excluders removed.

While indicating that the upper colony could raise its own queen or that a good queen cell – or a mailed caged queen – could be used, Farrar strongly advocated using bees with a new queen in a well-established a nucleus colony.

This technique establishes and builds the upper unit quickly. At the same time he under-supered the new top unit with two empty supers of drawn comb to facilitate expansion of the founding lower unit. Once established, the two units were united by replacing the dividing inner cover with a queen excluder, with additional supers also being added to the upper second-queen unit as needed. Farrar noted that the system – in some later developments – should be able to employ about half the number of shallow supers using full depth supers.

From here on in brood chambers were reversed regularly – until the colony got too large – and honey supers were under-supered as the season progressed to maximise brood production and to facilitate ready removal of finished supers of honey. Honey was harvested regularly, since supers fill very quickly in two-queen system operations while special care is required to always ensure that queens are kept quite separate during hive manipulations. The option of using an additional queen excluder above the second (upper) queen, though while not needed, will assist in keeping track of queens in extremely large and populous hives. Towards the end of the main honey flow all excluders were removed and honey supers were progressively removed to facilitate reversion of colonies to strong single-queen colonies to overwinter.

His late April 1954 paper reiterates his old scheme but emphasises that the central honey supers should not be allowed to fill as this would effectively divide the colony so that it would then operate as two single-queen colonies.

The sophistication of his operation is given brief insight to in one of Eva Crane's earlier visits to America[174]:

> We visited the apiary which Dr Farrar uses for his methods of management' experiments, which have recently been described in detail in Bee World (see Farrar, 1953). I had read and heard much about these colonies; even so, I was impressed by what I saw. The colonies, with 80-90 000 bees, were in shallow 12-frame Dadant bodies, 8-9 per hive. The two-queen units had already been united when we saw them. When the hives are to be inspected, a truck drives between the rows, and each hive in turn is lifted and turned over sideways on to a platform, and examined while on its side.

A surviving 1953 photograph (Figure 2.9) of this famous apiary is recorded in a report of an interview Clayton Farrar[175] had with Bud Cale where he asked:

Isn't your two-queen system of management too much work?

To which Farrar replied dryly:

If you are interested in getting those kind of crops, you will find a way to adapt yourself to the management.

It is worth noting that Cale, also a keen two-queen hive practitioner, had reservations about when or otherwise to operate two-queen colonies[176]:

One thing we have found out about the two-queen system is that not all colonies are candidates for two queens...

This year, only about a third of my bees are candidates for two-queen management [many were not strong enough], and the rest just won't be up to that sort of management.

Figure 2.9   Farrar's two-queen apiary referred to by both Cale and Eva Crane

The emergence of long idea hives, 20-35 frame horizontal Langstroth hives, raised the possibility that they might be employed to run two-queen colonies[177]. Long hives, as opposed to tall tiered vertical stack hives, date back to their invention by General Adair[178] in 1872-1873 and their subsequent popularisation by the military veteran O.O. Poppleton. The pages of *Gleanings in Bee Culture* and the *American Bee Journal* were awash with their use around 1872-1873[179] but, due to their unwieldy nature and often claimed failure to overwinter well[180] – Poppleton's warm climate operations excepted – their use soon fell into disfavour.

There was renewed interest in their use around 1900[181] by which time they had assumed a variety of names: Long Ideal Hive, Long Idea Hive, New Idea Hive... Muth-Rasmussen[182], in raising their confused status, drew clarification in a direct response from Poppleton:

> *Last summer, while looking over some copies of an Australian bee-paper, I noticed the name it used was Long-Ideal Hive, and I recognized at once that this was the solution of the name question. This name is so similar to the old one as to create no confusion whatever in making the change, is short, and, above all other considerations, violates no rules of good taste. The name is in common use in Australia, and, if I have any influence with the editors of our American bee-periodicals, it will be the common name here in America.*

> *While in Philadelphia last fall, I requested Editors York and Root to use the word Ideal instead of Idea when naming the hive, and as the most extensive user of these hives in this country (which, according to Mr Muth-Rasmussen's statement, are not the same hives to which the old name was first applied), I think my judgment as to which is the best name for the hive should have some weight. O.O. Poppleton, Dade Co., Fla.*

By around 1920 the use of long hives had gained considerable resurgence. For example E.R. Root[183] employed fourteen-frame hives, divided centrally by a vertical queen excluder, for queen rearing purposes. Others, such as Reeder[184], employed long hives on a large scale for comb honey production. Many writers, not least the influential Root family and publisher of *Gleanings in Bee Culture*, extolled their virtues in avoiding the handling of heavy supers.

Then in 1937 John Eckert[185] promoted the notion of horizontal two-queen hives designed to overcome the complex supering and brood manipulation requirements of Farrar tower hives. Eckert reported on two designs, a twenty four frame setup formulated by W. Leyroy Bell, Ralph Benton and Eugene Baker and a more standard fourteen frame hive adopted by Samuel Lawrence.

The duo-queen Bell horizontal hive comprised two seven-frame brood nests divided from a central ten-frame honey and pollen storage compartment by two queen excluder panels (Figure 2.10). Gracia Bell[186] reports that Leyroy Bell had named it the Bell Duo-Hive. There was a main central entrance to the ten-frame compartment and small entrances for each lateral brood chamber. No substantive details of the operation of the Samuel Lawrence horizontal hive were recorded by Eckert.

In the view of one eminent group of beekeepers[187], two-queen long hives are both difficult to handle and maintain.

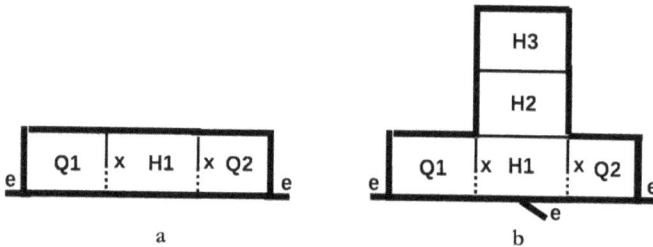

Figure 2.10 Bell horizontal duo-queen hive (after Eckert): e = entrance; x = excluder; Q = queen; H = honey super:
 (a) flanking brood nest set up; and
 (b) supering for the flow, central entrance opened for the honey flow.

Enhancements to the original Farrar two-queen hive

By the late 1930s Charles Harold Gilbert[188], and indeed also Farrar, had explored many top-bottom two-queen tiered hive arrangements, for example placing weak overwintered colonies over a screened inner cover, introducing a second queen to upper units as a requeening ploy and using two-queen setups to control swarming. Somewhat later Roberts and Mackenson adapted the top-bottom Farrar two-queen style hive for queen raising using the lower colony to boost the colony while maintaining the breeder queen unit on top[189]. A ripe queen cell introduced to young brood over a queen excluder rather than a double screen, well separated from the main brood nest, resulted in reliable establishment of a second queen[190].

Gilbert's modification of the Farrar system (Figure 2.11) employed standard ten frame full depth Langstroth hive bodies and substituted the inner cover with a double screen. Use of a double screen in lieu of an inner cover allowed for more effective warming of the upper embryonic colony. Initial separation of queens during an establishment phase is essential to avoid loss of one of the queens[191].

In Gilbert's scheme the double screen was replaced with an excluder after several weeks once both queens were laying well. He also employed a riser rim over the excluder with an opening and an attached landing board obviating the need to drill auger holes to provide the upper brood nest free flight access.

In referring to Eckert's and Farrar's experience of increasing bee populations using two queens, Gilbert noted that:

> Success is based upon greatly increased bee population resulting from the combined efforts of two queens working in the same hive. The method may appear new and revolutionary to some beekeepers but, according to Eckert its use dates back many years in the history of beekeeping.

He also noted preemptively:

> ... that two-queen colonies require regular manipulation, and that serious difficulties arise when they were neglected. Supering requirements increase greatly during a heavy honey flow, and it was almost impossible to give each unit enough supers.

Gilbert's use of double brood boxes reflects the common American practice of running two brood boxes in single-queen hives. While a single full depth nine or ten frame brood chamber (or an eight frame full depth super combined with a shallow or ideal super) normally provides adequate space for prolific queens to lay, there remain many adherents to double brood chambers.

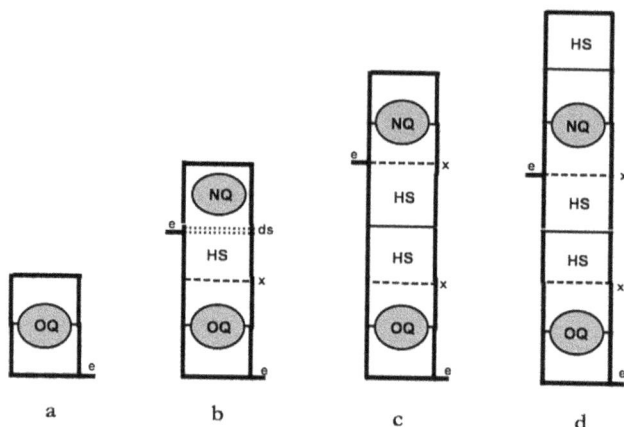

Figure 2.11   Charles Gilbert's system operation of two-queen colonies for ten-frame shallow super operation: e = entrance; ds = double screen; x = excluder; OQ = old queen; NQ = new queen; HS = honey super:

(a) a strong single-queen colony is established using a good previous season queen and is built early in the season;

(b) the colony is split, a new queen is introduced to the upper unit above a double screen and the lower unit is supered above the excluder to facilitate expansion;

(c) the colony is united and the top unit undersupered; and

(d) the two-queen unit is further supered and dynamically operated during the honey flow.

Winston Dunham's modified system for honey production[192] developed from the late 1930s was a further adaptation of the Farrar two-queen system. Instead of maintaining two queens and supering each of the brood nests separately in tower hives – a labour intensive element throughout extended honey flows – hives were reorganised and reduced to a single-queen condition (Figure 2.12). The scheme, trialed over nine years and tested with commercial producers, targeted flows of limited duration, such as the clover flow in the US State of Ohio that lasts for 6-8 weeks. It would seem to have particular application to Australian box eucalypt flows that are equally well defined.

The arguments put forward by Dunham, for example the need for young and productive queens and contingency for managing colonies of different strengths and dynamic management of backup colonies, signals that there are natural limits to the number of such hives (Dunham suggested around 300 not counting support colonies) that can be effectively managed. Close reading of the condition of individual hives is essential and a very flexible approach to their operation is needed especially during a honey flow.

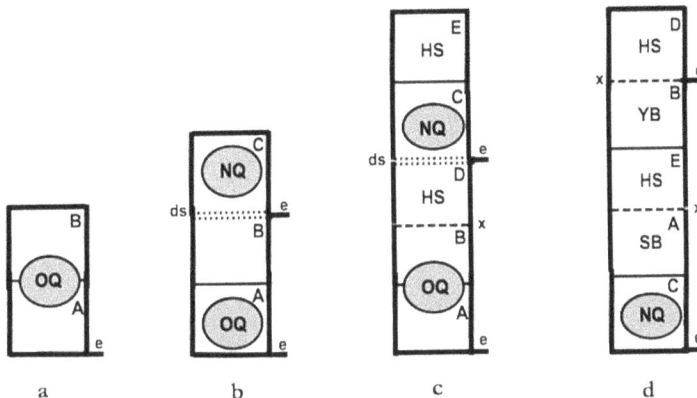

Figure 2.12 Winston Dunham's Ohio modified two-queen system: e = entrance; ds = double screen; x = excluder; OQ = old queen; NQ = new queen; HS = honey super; SB = sealed brood; YB = young brood:

(a) only medium to strong double hives are selected;

(b) three frames of sealed and emerging brood and three frames of honey are moved above a double screen and a caged queen is immediately introduced;

(c) each brood nest is supered to facilitate buildup of both brood nests; and at the beginning of the main flow; and

(d) the new queen is moved to the bottom board to facilitate colony requeening while the old queen is removed and brood from the old brood nest is reorganised so that emerging sealed brood is located above the new queen. Frames with eggs and young brood are located further up the stack to encourage bees to stay in the honey supers.

Similar detail is found in Dunham's 1947 Ohio Extension Service Bulletin report and allied publications especially in terms in terms of strategic timing of hive operations. Eckert and Shaw[193] in reviewing Dunham's 1943 publication noted that he:

> ...set up his two-queen unit in much the same way as did Farrar, but instead of a regular inner cover he used a double-screened divider with a quarter-inch of space between the two screens. The full screen allows more heat from below and aids in the development of the weaker colony above...

and that:

> ...the goal sought in Dunham's system of management of two-queen colonies is to reach the desired strength before the main honey flow while the equipment is still light in weight. There must be a continuous light nectar flow to provide adequate stores and uninterrupted brood rearing. By the end of the spring breeding period, the ideal two-queen colony will have 10 to 14 frames of brood in the lower unit, 8 to 10 frames in the upper unit, and the hive crowded with bees.

As Dunham surmised:

First, it is essential to understand the difference between the standard two-queen system and the Ohio modified two-queen system. The standard two-queen system utilizes two queens at least during a part of the building-up period and throughout the harvest period. During the harvest period supering involves going through the colonies every 10 days and supering the lower and upper units, which are each headed by a queen. This system is best adapted for a region characterized by a long honey flow, and represents an intensive type of beekeeping, where maximum yields of honey are harvested.

But Dunham had other insights into the value of setting up two-queen colonies. In an earlier 1943 paper, he addressed practical means of bringing all of his many hives up to full strength by the beginning of the main flow. To do this he established ten to twenty percent of his colonies as two-queeners. These he used these to recover all hives that were below par providing explicit instructions on how to remediate specific problem conditions, namely:

- colonies that were in a weak-to-medium condition, but normal in other respects;
- colonies that were weak-to-medium in strength but which possessed an old, failing, inferior queen, or which had become queenless for a short time;
- strong colonies in a queenless condition or in possession of an old, failing, inferior queen;
- colonies that were of normal-to-medium strength but were backward in occupying supers; and
- strong colonies with a normal amount of brood but showing dominant swarming characteristics.

Eckert and Shaw in reviewing the wider apiary scene noted that:

> Some beekeepers operate two-queen colonies in specially built hives, some of which have the queen confined to brood compartments on the sides with storage chambers in the center. Supers are added to the center stack as they are needed during the season. Some difficulty is generally encountered in keeping honey-free brood chambers and enough room for good queens to maintain a high rate of brood rearing. Such hives are too heavy to move where migratory beekeeping is practiced but could be used to advantage in permanent apiaries. Most beekeepers, however, prefer to use standard hive bodies because they are interchangeable and more easily managed.

Then at the end of the season Eckert and Shaw[194] found a singular new use for strong colonies to overwinter new queens:

> In the more temperate regions, weaker colonies with good queens can be wintered above a stronger colony if separated from it by a double wire-screened division board or a thin liner cover with the bee escape covered with a double wire screen. In fact, in some areas many commercial beekeepers divide their stronger colonies six to eight weeks before the close of the brood-rearing period and place the divide over the parent colony, separating it with double screen or a thin division board. The upper colonies are always given a separate entrance. The divides are given young queens or, if time permits, ripe queen cells. The divisions should consist of five or six combs of brood and bees and enough honey to winter on, unless both colonies can store sufficient honey for this purpose.

In a contemporary variant of the Farrer system, Victor Croker and David Leemhuis at Australian Honeybee introduce ripe queen cells raised from select stock to several frames of young wet brood lifted above an excluder much as Roberts and Mackenson had done. This allows each nucleus colony on top of each stack to quickly establish new queens by provisioning each fledgling broodnest with stores from below and – as we have often noted – to benefit from the rising warmth of the brood with the old queen below. Australian Honey Bee do not run the resultant two-queen colonies as production hives, instead employing the newly raised queens to directly requeen each hive. In their operation, and once the new queens are well established, the brood chambers with the old queens are simply offset to new stands to allow field bees in the offset hive to drift back to their parent stands. At the same time each top unit with a new queen is moved to the bottom of the stack. This provisions every hive in the apiary with a new queen and leaves all colonies in optimal condition.

In the Australian Honeybee operation, the newly offset hives – with the old queens and depleted of field bees – are moved to another apiary where they are readily searched and made ready to accept a new batch of cells.

Clearly any scheme that facilitates hive renewal and that minimises the amount of labour required is well worth pursuing, many capable operators employing a second queen, in one way or another, to avoid honey bee colonies falling below par. Wedmore's 1945 classic book, *A Manual of Beekeeping for English-speaking Bee-keepers*[195] reviews nearly 1600 hive manipulation practices. He makes a specific recommendation about employing young queens, about using the similar Demaree plan to establish new queens and to strengthen swarms and weak colonies with good queens in tiered top-bottom two-queen systems. He draws conclusions not dissimilar to those made by Gilbert and by Eckert and Shaw.

Having placed the two-queen hive on a sound practical footing, we can chronicle the post 1950 endeavours of the likes of New Zealand researcher G.M. Walton and the renowned American apiarist Robert Banker to demonstrate the commercial viability of large-scale two-queen hive apiary operations.

IV– Commercial Two-Queen Hives

My son, eat thou honey, for it is good.

Proverbs 24:13

The post Second World War II era saw a period of successful operation of commercial two-queen hive apiaries. We encounter the likes of Robert Banker (an editor of the beekeeping bible *The Hive and the Honey Bee*), New Zealand's Department of Agriculture and Fisheries' inventive G.M. Walton and Don Peer, a disciple of Farrar. They led the way in defining the limits of maintaining a second queen and in devising timely ways to harvest enormous honey crops. Until these beekeepers arrived, the large-scale operation of two-queen hives had been an elusive exercise, their very operation being the territory of the most skilled apiarist.

Many advances in the operation of two-queen hives were made in the post war period many incorporating an element of swarm control in their setup. Notable contributors included those of Holzberlein[196] (1952-1955), Miller and Cale[197] (1953-1954), Latif and coworkers[198] (1955-1960), WaFa (1956)[199] and Haydak and Dietz[200] (1967). Progressive overviews of two-queen systems are provided in the March 1953[201] and April 1954[202] issues of the *American Bee Journal*.

John Holzberlein's two-queen colonies

John Holzberlein, like his contemporary Winston Dunham, was a leader in efficient two-queen management practice. Their schemes greatly simplified hive management facilitating a golden era for their use in commercial operations.

Holzberlein identified two failsafe ways to arrest swarming[203]:

> *...One is to swarm them, that is, some phase of dividing the old bees and queen from the young bees and brood. The other is to make the colony queenless and queen cell-less, causing it to raise a young queen from scratch.*

Alone neither, he concedes, is feasible in the commercial operation as both are counterproductive to building bees for the honey flow.

Holzberlein's ingenious solution was to do both, founded in a lesson that he learnt from the Nebraskan beekeeper Ralph Barnes, of Oaklands whose mantra was to divide or requeen:

> *The kind of dividing I am going to tell you about has no part in making increase. The divide is made all under one cover. The split or divide is set up*

over a solid or screened inner cover with an entrance of its own, and given a young queen. It is the beginning of the two-queen system, but right now it is a divide and the best little swarm preventer that you ever tried. Aside from being almost sure-fire swarm prevention it has the added advantage of getting and holding more bees in the field force of the colony than one queen could possibly produce. It keeps them all coming to the same hive, yet divides them at a time when the desire to swarm is almost sure to take over if nothing is done...

So saying, Holzberlein outlined the details of his technique (Figure 2.13)[204]. The original setup is redrawn to depict now standard shim entrances, rather than the original auger-hole exits, and to show the location of queens but I recommend close reading of his original articles to gain a full appreciation of the subtleness of his scheme. In reflecting on the use of a second queen to build enormous colonies to take full advantage of major honey flows[205] he notes:

The two-queen system of management is not much different from standard good practices of management where only one queen is used. The colony must have had an abundance of overwintered stores, a large cluster, and a good queen. Pollen must be supplied in some form to ensure early spring brood rearing, for if the colony is to be operated as a two-queen unit it must be a super colony, the kind that would be almost sure to swarm if it were not divided. There is no use to waste time on weaklings. We have used this system long enough to have worked out most of the details and cut some corners.

The elegance of Holzberlein's scheme lies in the great simplicity of its setup: he could set up one hundred colonies with two queens in a day. He emphasised the need to only use strong colonies and the value of first setting back these colonies and placing a new bottom board and empty super on the parent hive stand to facilitate ready sorting of frames.

Into this empty chamber he placed unsealed brood and some stores, placing five of six frames of sealed brood with emerging bees either above an excluder, if he failed to see the queen, or above a double screen or inner cover, if the queen was found, when he then placed the colony queen in the bottom brood chamber (Figure 2.13a and 2.13b). Once settled a new queen was introduced above a double screen and the brood nests supered to build bees for the flow.

Then at the beginning of the main clover honey flow he united the upper and lower units obviating the need to find the founding queen. In doing this he avoided the complex manipulations propounded by Farrar for operating two-queen colonies through the honey flow.

In surmising the operation of two-queen hives, John Holzberlein[206] noted saliently:

> For some 14 years we have been using two-queen management with varying degrees of success. It is like any other method of beekeeping in that the answers are not all apparent at once, nor will any set of rules work all of the time. Some basic principals always apply, however, and the details can be worked out to suit one's local conditions.
>
> ... The extensive producer with widely scattered yards and time for only infrequent visits will not find the two-queen system suitable to his system. Nor will the migratory man who picks his crop in two or more locations. But the producer with limited territory and reasonably certain flows may well increase his production, cut down his losses from swarming, and yet operate no more colonies nor occupy any greater territory. The amount of extra work necessary is in very favorable ratio to the extra amount of honey secured, thus lowering the cost of production per pound of his crop.

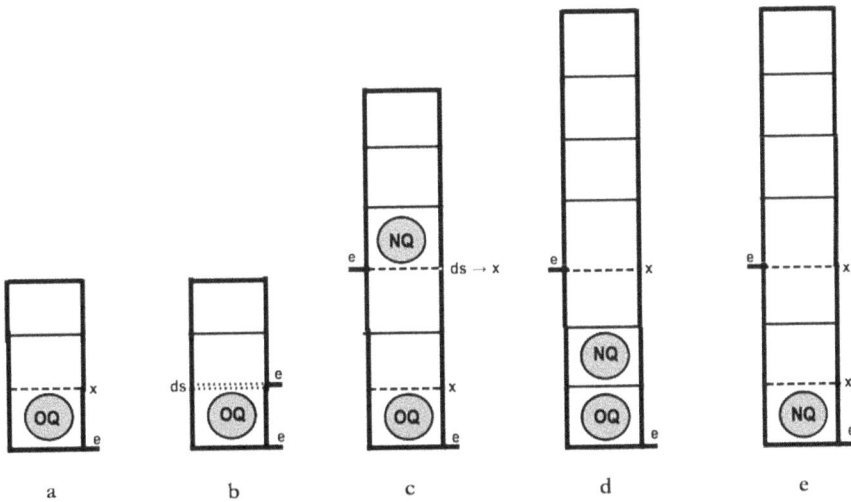

Figure 2.13 Holzberlein's setup of a two-queen colony: e= entrance; ds = double screen; x = excluder; OQ = old queen; NQ = new queen:
The first stage is set up based on whether the queen is seen during initial frame sorting:
(a) if the queen is not seen, any bees not already in the bottom brood chamber are shaken in front of the hive. Worker bees pass through the excluder as needed to attend any brood present above the excluder leaving the queen below; thereafter the excluder is simply replaced with a double screen (13a → 13b); or
(b) if the queen is seen, she is simply located in the bottom brood chamber

below a double screen (or inner cover or nucleus board) (13b). Ample young bees are shaken off brood frames into a super above the double screen. Older bees will drift to the bottom brood chamber, but nurse bees will remain with any brood present. In either instance the queen will be found in the bottom brood chamber;

(c) after a short time, when the hive is settled and the bees are fairly evenly distributed, the colony is reorganised and a caged queen is introduced to the top chamber together with five to six frames of sealed and emerging brood and abundant nurse bees. Once the new queen is laying well, the double screen is replaced with a queen excluder – as shown – and supered for the honey flow;

(d) near the commencement of the honey flow, the brood chambers are juxtaposed – not shown in original drawings – allowing the usually fitter young queen to supersede the older queen; and

(e) the colony is organised for the honey flow as a single-queen hive.

Schaefer outlined a very similar scheme[207] employing caged queens installed at the time of make up of the vertical split or alternatively introducing queen cells.

Robert Banker's and Don Peer's two-queen colonies

By the late 1960s apiarists Robert Banker and Don Peer were operating very large apiaries based on Farrar's two-queen system. Banker[208] operated 1500-1600 two-queen hives with an additional 1000-1200 packages and divisions, a remarkable operation. In a very sophisticated and highly nuanced scheme he was able to maintain all production two-queen colonies at maximum strength throughout honey flows, while also averting swarming.

Banker made dead-of-winter preparations to ensure that absolutely everything was ready for the bee season. By early spring he commenced building strong double brood colonies, first lifting brood and reversing lower brood chambers to ensure he had strong colonies prior to dividing them to establish an extra queen.

The timing of introducing new caged queens to the top unit was critical to the process of maximising colony strength. He ensured that colonies with two queens had 6-7 weeks to build for the honey flow. A defining feature of Banker's operation was contingency for every condition, ensuring starting queens were performing satisfactorily, strengthening colonies with brood, feeding bees sugar and pollen in the buildup phase, conducting rigorous brood disease checks and ensuring all bees were at optimal strength both at the commencement and right through the honey flow.

Banker employed ten frame gear throughout, full depth supers for brood chambers and shallows for honey supers, the latter for ease of handling and removal of the crop (Figure 2.14).

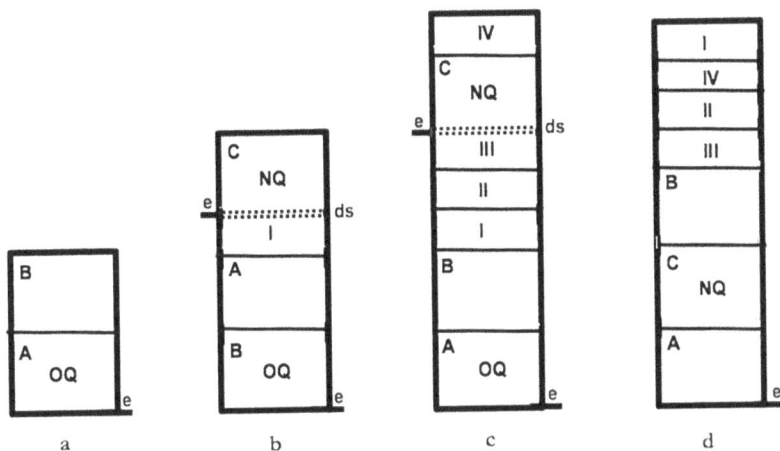

Figure 2.14   Robert Banker's 1968-1979 system for operation of two-queen colonies for ten-frame shallow super operation: e = entrance; ds = double screen; OQ = old queen; NQ = new queen:

(a) a strong double is built early in the season using a good previous season queen or replacement queen if she were not performing optimally;

(b) the double brood nest is reversed with roughly three frames of brood with adhering bees moved above an excluder to draw nurse bees up. Shortly thereafter the excluder is replaced with a double screen and a new queen is introduced and established in the upper unit. The lower unit is supered above the lower double brood chamber – and below the double screen – to allow expansion of the lower colony to avoid swarming;

(c) the two-queen unit is further supered on a rotating super basis during the honey flow; and

(d) the brood and super chambers are rearranged towards end of the honey flow to allow supersedure of the old queen[209] and finally returned to the double brood chamber condition (d → a) to overwinter.

It is worth noting that the old queen may survive though this is probably less common and very occasionally both queens may be lost. Note also that the system was operated with a double full depth brood chamber for the lower queen and a single full depth brood box for the upper queen. With most brood occupying the respective brood full depth brood chambers, the queens were never restricted in their egg-laying capacity.

Robert Banker's and Don Peer's systems represented the apogee of development of two-queen hives as best understood in the 1960s and 1970s. Eva Crane[210] reports on two large scale apiary two-queen enterprises, Don Peer's on the Saskatchewan River and that of Mr and Mrs Warren at Babe's Honey Farm on Vancouver Island in Canada, using package bees. Of Dr Peer's operation[211] she reported:

> *Don Peer uses two-queen colonies on Dr Farrar's system... He gets 2-lb. packages about 23rd April, and uses two per hive, the division board between them being later replaced by a queen excluder. By 23rd June or so the 16 000 bees in the two packages have produced 70 000 or 80 000; by July there are 90 000 or 100 000 in each hive. A thousand or more hives, in apiaries of 25- 35, are spread over an area perhaps 50x30 miles, parts of which are also used by other beekeepers.*

Pointedly Eva Crane also noted:

> *Don Peer left his career in bee research several years ago, because he could get a much higher income by producing honey.*

Walton's two-queen hives

At the same time during the 1970s, New Zealander Walton[212], working with apiarist Dudley Ward, conducted a large study comparing the performance of single-queen and two-queen hives. In their trial, strong double colonies were reversed then supered to facilitate expansion and to more evenly distribute brood and stores. The single-queen hives were operated in the normal way employing a double brood box (Figure 2.15).

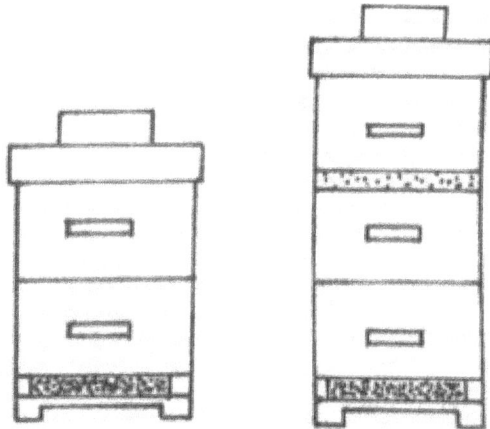

Figure 2.15 Walton's 1974 brood nest setup comparing performance of commercial single and two-queen colonies:
(a) single-queen hive; and
(b) two-queen hive.

For the two-queen Walton setup, a subsequent split was employed to establish the second queen, a mailing cage queen being introduced above a double screen. After 4-6 weeks the double screen was replaced with a queen excluder, each queen occupying a double brood chamber. Subsequent restriction of the upper queen to a single brood chamber was employed to encourage honey production as a tradeoff for loss of some worker bee production (Figure 2.16).

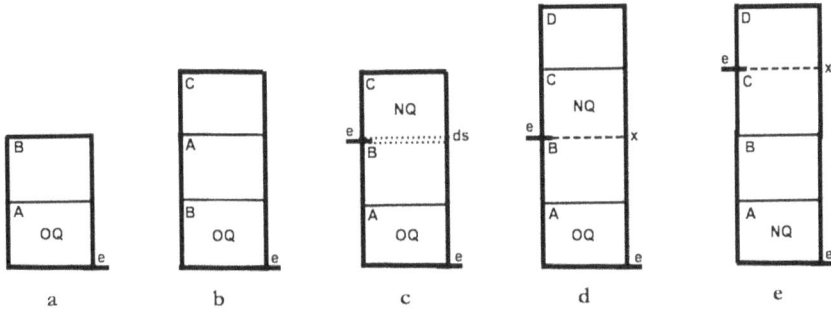

Figure 2.16 Walton's system for operating two-queen colonies: e = entrance; ds = double screen; x = excluder; OQ = old queen; NQ = new queen:
(a) a strong double is built early in season employing a strong overwintered hive with an old queen;
(b) the double brood nest is reversed and supered;
(c) a two-queen unit is established by introducing a new queen above a double screen;
(d) the colony is united by replacing double screen by an excluder at the commencement of honey flow; and
(e) the excluder is removed towards the end of honey flow and to allow supersedure of the old queen.
Note: Supers are progressively removed at the tail of the flow returning the hive to a strong double for overwintering (e → a).

The enduring legacy of Walton's study was that of clearly demonstrating the superior honey gathering performance of two-queen colonies.

Nabors[213] reviews the main findings of two-queen colony operation reported by Farrar, Peer and Walton. He not only concluded that two-queen colonies greatly increase honey production but also compared details the different styles of operation of two-queen hives. In reviewing the setup of two queen hives, Nabors contemporary Hesbach[214] noted that:

...two-queen systems [have] almost always [been] configured vertically but have since been configured in both vertical and horizontal systems; and

89

> *...vertical systems are upright stacks with one brood chamber on the bottom board followed by honey supers, a queen excluder, and another queen-right brood chamber on top.*

While these conclusions ignore the consolidated brood nest findings of two-queen systems formulated by John Hogg and Floyd Moeller, the new systems were simpler to operate than required in the original Farrar plan.

In a further refinement of their system, one that Canberra Region Beekeepers have been experimenting with for several years, Bernhardt Heuvel[215] records a local professional beekeeper employing a jumbo bottom brood super with a central vertical excluder, queens to each side, claiming it produced more honey and less work than the Farrar-style two-queen set up. The horizontal configuration certainly avoided honey being stored in the upper brood chamber, but there were, nevertheless downsides. These include the upper brood nest not benefitting from the rising warmth of the lower brood chamber and the use of a non-standard oversized lower double brood chamber.

V – The Consolidated Brood Nest
Two-Queen Hive

The queen bee never stings unless she has such an advantage in the combat that she can curve her body under that of her rival in such a manner as to inflict a deadly wound without any risk of being stung herself!

The moment that the position of the two combatants is such that neither has the advantage, and that both are liable to perish, they not only refuse to sting, but disengage themselves, and suspend their conflict for a short time!

L.L Langstroth
The Hive and the Honeybee 1853[216]

So far we have described various iterations of the Farrar plan of two-queen hives, piggybacked single-queen colonies. Each brood chamber had its own entrance and honey supers. Early in the season the upper unit benefited from the resources of the brood box below during the establishment of a second queen while, as time passed, the whole colony benefited from the combined laying power of two queens. As with doubled hives, developed by Wells in the early 1890s and two-queen hives Farrar had established, the combined workforce not only operates cooperatively but also a proportionally larger number of bees are available to harvest nectar efficiently. So what further improvement could be made to systems with two queens?

The consolidated brood nest two-queen hive

An enormous advance in the operation of two-queen hives was the invention of the *consolidated brood nest* (CBN) hive. The CBN scheme replicates conventional single-queen hive management, brood below, honey supers on top. In so doing, Floyd Moeller and John Hogg truly revolutionised the running of two-queen hives.

Floyd Moeller[217] was the first researcher to recognise that complete separation of brood nests, each with its own queen, was quite unnecessary, his original setup (Figure 2.17) clearly illustrating the change in their configuration. As he was keen to point out:

Most of the methods tend to emphasize separating brood nests, presumably to prevent queens from fighting across the queen excluder. This is no longer considered critical. Some earlier methods are rather complex and unnecessarily involved. The two-queen system described here is simple and still uses the normal behavior patterns of bees.

This was a critical inflection point in the long road to efficient contemporary two-queen hive operation. Moeller and Hogg introduced a range of innovative hive queening techniques that are now helping address problems such as declining honey yields and increased disease prevalence.

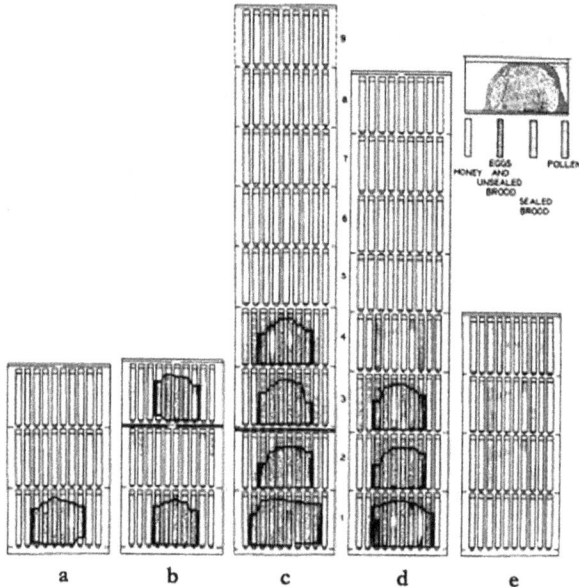

Figure 2.17   Floyd Moeller's original 1976 CBN two-queen hive:
(a) the process starts with a strong starter colony;
(b) brood lifted is lifted above double screen and a new queen is established;
(c) the colony is further built (the double screen being replaced with a queen excluder) with no restriction on space for each queen to lay and is top supered to take the main crop;
(d) the excluder is removed near the peak of the honey flow; and
(e) the colony is collapsed towards the end of the honey flow.

With this growing understanding, Floyd Moeller[218] operated two-queen colonies with queen excluders juxtaposing their brood nests. Moeller employed either standard full depth ten-frame gear or shallow twelve-frame gear throughout a six-year 1967-1974 study, shown here for his ten-frame operation (Figure 2.18).

To establish his two-queen hives, brood chambers were reversed from early spring to stimulate brood rearing. Once each colony grew to a strong double the colony was split and a caged queen was introduced to whichever portion of the split colony did not have fresh eggs after several days. This obviated the need to find the parent colony queen, an obvious advantage in any large scale commercial operation.

Initially a double screen was employed to separate the piggybacked splits, the upper colony benefitting as usual from the rising warmth of the brood nest below. With this arrangement, and once well established, the two units could be united by the simple measure of replacing the double screen by an excluder.

Moeller's system was very similar to that of Walton except in that ample space was always provided for brood rearing at all times, the upper queen only limited by her laying capacity.

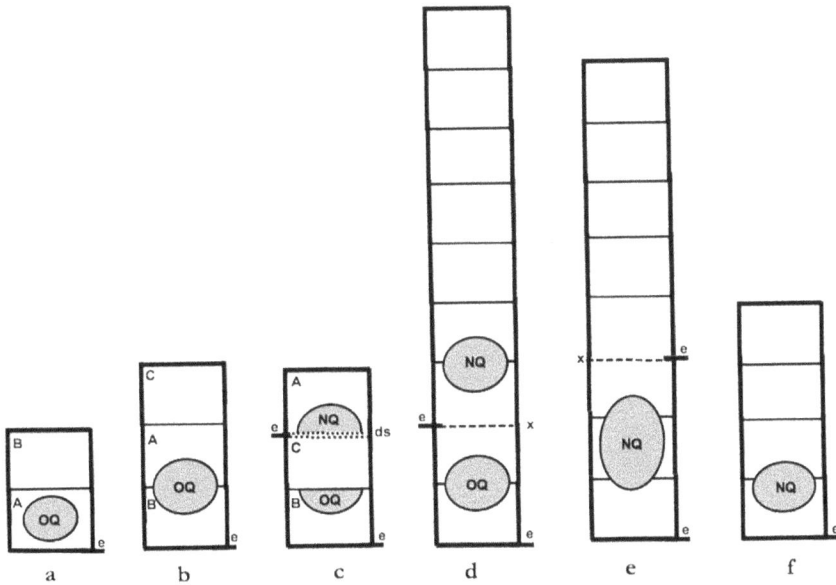

Figure 2.18 Floyd Moeller's system for operation of two-queen colonies, redrawn and modified to show operational detail: e = entrance; ds = double screen; x = excluder; OQ = old queen; NQ = new queen:
(a) the process is started with a strong overwintered colony with an old queen;
(b) the colony is built to a strong triple;
(c) the colony is split and a new queen is introduced above a double screen;
(d) the colony is further built with the double screen replaced by a queen excluder giving both queens given ample room to expand and enable simple top supering for the honey flow;
(e) the excluder is removed towards the end of the honey flow facilitating staged removal of honey supers. This also allows the top younger queen to supersede the old parent hive queen; and
(f) the colony is further collapsed in preparation for overwintering (e → f → a).

Towards the end of the main flow, the colony was collapsed to a single-queen hive this time achieved by removing the excluder dividing the large brood nest. Here any effort to extend the capacity of the colony to raise more workers would be wasted and a drain on colony resources. Timing of colony building to coincide with major honey flows is important in any operation but regulation of population size is even more critical for operation to two-queen colonies.

Moeller stated the obvious by noting that, under conditions where a single-queen colony required a single honey super, a two-queen colony required at least two supers. In comparing his ten-frame full depth system with that of his twelve-frame shallow super operation, he observed that shallow supers were filled more quickly and were thus more easily removed and returned to the hive. Rapid honey removal is a key feature of any honey flow management program but is essential in two-queen hive operation.

Moeller's lasting contribution was the abandonment of the complex Farrar-style supering arrangements. All honey supers are operated on the single-queen colony plan: all honey supers were located above the brood nest. Moeller later compared the single-queen and two-queen systems of management and concluded that well-run two-queen hives maximised honey production[219].

The rationale for and means of establishing and operating colonies with two queens in a single brood nest have been fully canvassed elsewhere[220]. In a major refinement of the double brood nest approach to the operation two-queen colonies, Hogg[221] formulated sound theoretical principles – and a range of practical measures – for establishing a single consolidated brood nest (CBN) containing two or, in principle, many queens. He first investigated the full conditions for queen acceptance to any honey bee colony[222]. His essential finding was that the condition of acceptance of two or more queens were essentially the same as those for acceptance of one queen. A colony made gyne free – that is without any queen or potential queen – will, depending on the setup and prevailing conditions, readily accept two or more queens.

Here, as conceptualised by Hogg, instead of independent brood chambers, queens share in the development of a single but extremely large brood nest. Pheromone signalling and brood raising are fully integrated lending intrinsic stability to a colony so large that, in the absence of a honey flow would be very swarm prone[223]. Since active brood nest temperatures hover around 35 ^{0}C, they operate mainly above ambient air temperature. And since the thermal efficiency of a brood nest improves as the nest expands by a factor of 4/3 times the radius*, the two-queen broodnest is more efficient bee-for-bee than in the much smaller single-queen colony. While two-queen hives may benefit from other economies of scale, such as more efficient scouting for nectar, pollen, propolis and water and distributed decision making may be improved, quantifying and pinpointing the importance of each of these factors is not understood.

*The volume to surface area ratio is four thirds times the radius assuming a spherical brood nest:

$$V = \frac{4}{3}\pi r^3 \qquad A = \pi r^2$$

Hogg concluded that the new arrangement (Figure 2.19) would provide the basis for new and less complicated two-queen management systems, especially applicable to comb honey production[224]. He suggested that two-queen colonies generate 25-50 % more bees than does a single-queen colony and were anyway less swarm prone.

Hogg's very elegant two-queen scheme culminated in the ingenious Juniper Hill Plan[225] designed not only to minimise swarming and to increase honey production, but also to revitalise the art of comb honey production.

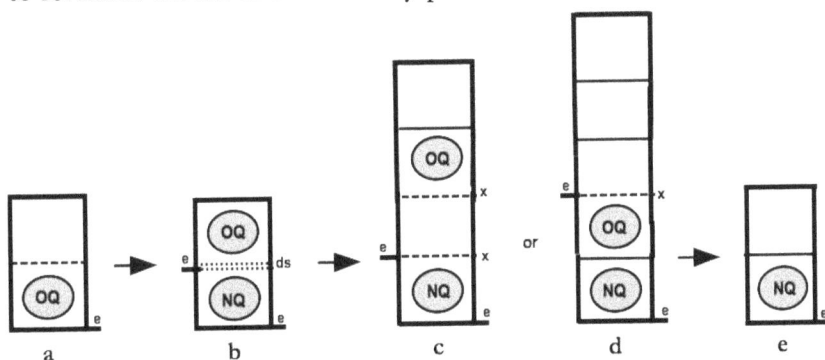

Figure 2.19   John Hogg's consolidated brood nest (CBN) model for operation of two-queen colonies: e = entrance; ds = double screen; x = excluder; OQ = old queen; NQ =new queen:
(a) an overwintered colony with old queen is built to a strong double in early spring;
(b) the colony is split using a double screen and is queened; and, once well established, either:
(c) the double screen is replaced with a queen excluder (employing single 10 or 12 frame brood supers), and is supered for the flow, or
(d) the double screen is replaced with a queen excluder, employing an expanded bottom brood chamber and a single upper brood chamber (for eight-frame operation) and supered for the flow; and
(e) the queen excluder is removed towards the end of the flow when the colony then reverts to a strong single-queen unit for overwintering.

In his later writings, Hogg contemplated the exigencies of two-queen hives that might either swarm or where supersedure might occur. In the process he posed questions about our understanding of the operation of colonies with more than a single queen:

> *When swarming occurs, will one or both queens go out with the swarm; and will queen succession occur accordingly to maintain double queen status? Also, will supersedure of one or both queens ever occur.*

Interestingly Farrar[226] touches on this issue in observations he made on supersedure in two-queen colonies noting that:

> *The problem of supersedure is no greater in two-queen than in single-queen colonies except that, when one queen is poor and queen cells are started, cells usually will be built in the other brood nest.*

Dannielle Harden, from Canberra Region Beekeepers, and I have made the same observation. One queen from a two-queen hive was lost under then poor seasonal conditions of 2019-2020. Queen cells appeared in the queenright portion of the hive in the same manner reported by Farrar. There was however no worker or drone brood of any age anywhere in the hive but, on a subsequent inspection four weeks later, a new supersedure queen was found to be laying well, the colony having benefited overall from two laying queens in the spring and early summer buildup phase.

Consolidated brood nest operations

Operating two-queen hives optimally requires considerable skill, particularly in respect of timing of operations and in their initial setup. They can only be operated during the main spring buildup through to near the end of major honey flows and are prone to failure in any time of dearth.

Well set up, however, they can be operated in much the same way as single-queen hives – if the CBN system is employed – except in that honey must be removed promptly if their productive potential is to be realised. Unlike single-queen hive systems all should employ at least one queen excluder to ensure that the queens are kept apart and colonies need always to be managed to minimise the risk of the colony reverting back to one queen. This often reported finding was confirmed in my experience of running two-queen colonies over the very poor seasons of 2018-2019 and 2019-2020. Of nine colonies, five reverted to single-queen hives during repeated stop-start honey flows. While a second queen was successfully restored in four of those colonies, the final loss of one queen occurred too late in the season to justify requeening. Only one of the nine colonies performed well.

While there is much to recommend in the CBN system (Figure 2.20a), the original Farrar scheme (Figure 2.20b) has special application in accelerating the development of nuclei and in establishment of new queens. Where a new queen is established and laying well the supers and brood boxes can be rearranged to configure the colony as the CBN system (Figure 2.20a) or the new queen and her brood nest can be placed on the bottom board and the old queen removed as a simple means to effect requeening.

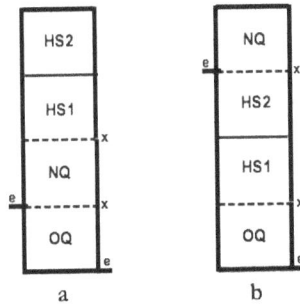

a   b

Figure 2.20   Basic vertical configurations of two-queen colonies: e = entrance; x = excluder; OQ = old queen; NQ = new queen; HS = honey super:
(a) Consolidated Brood Nest (CBN) with double brood nest with queens separated only by an excluder and supered above; and
(b) Farrar-style Hive with separated top and bottom brood nests, honey supers sandwiched between, if needed also supered above the top brood nest. The systems (as shown) assume ten or twelve frame super operation using standard full depth gear throughout. Many variants exist, a number employing shallow or ideal honey supers for ease of crop removal.

Two-queen hive configurations

There is also a clear recognition that two-queen hives can be configured vertically or horizontally while honey supers and barriers can be deployed in many arrays.

The more sophisticated juxtaposition of brood nest, the consolidated brood nest configuration (CBN), is reflected in the vertical and horizontal arrangements (Figure 2.21a and 2.21b). Where the two brood nests are kept discrete, the vertical configuration is reflected the top-bottom arrangement (Figure 2.21c) while the horizontal setup is reflected in the flanking arrangement (Figure 2.21d).

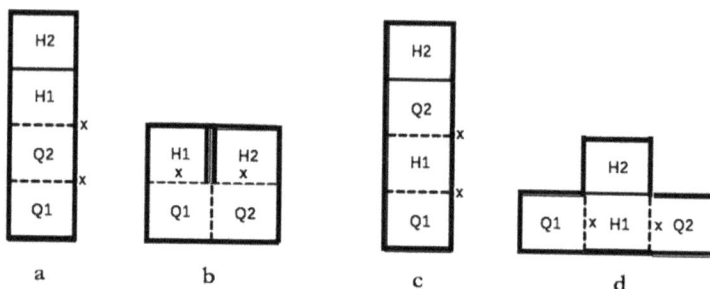

Figure 2.21 Vertical versus horizontal architecture of two-queen hives depicting brood-chamber and honey super alignments: x = excluder; Q = queen; H = honey super:

(a) CBN tiered arrangement;
(b) CBN side-by-side arrangement;
(c) Farrar tiered arrangement; and
(d) Bell duo side-by-side arrangement.

Recent applications of two-queen and doubled hive systems

All in all two-queen hive operation has been set on a very firm footing. While running successful two-queen demands intensive management, and a nuanced understanding of their operation, two-queen hives have also been used to address emerging problems of apiary operation such as declining floral resources and weakening of bees by disease.

The alternative being proposed by Tom Seeley, is to return beekeeping to a much more natural wild hive condition where colonies are much smaller and the returns (honey and pollination service) are much reduced but where bees, with some modicum of swarm and disease control, will survive and still provide traditional products such as wax and a small honey crop. In this context it is instructive to read the account of Eva Crane on the declining farming of traditional native stingless bees (Meliponini) in Mexico[227] where they are maintained as a commercial source of honey.

In the early 1990s, Duff and Furgala[228] compared the performance of 500 two-queen units made up from same-season package colonies, overwintered colonies (each hive with its division on the same stand) and parent colonies and their divisions hived on separate stands evaluating their productivity both in terms of management and their profitability (Figure 2.22). This study provides some insight into the most appropriate system for establishment of two-queen colonies tailored to factors such as latitude and floral phenology, the flowering pattern of plants. From around 2000, a number of North and Central American two-queen hives studies were undertaken to attempt to address the problems of local climatic extremes.

Figure 2.22   Duff and Furgala's 1990 single and two-queen hive study

Villarroel, Rebolledo and Aguilera[229] employed doubled and tripled colonies (rather than two-queen or three-queen hives) to investigate problems of declining honey production on the Mexican High Plateau attendant to the arrival of Africanised bees and *Varroa*. These authors describe a 44% honey yield increase with two queens. In another study on the Mexican High Plateau, also undertaken to address problems associated with declining honey bee yields, Africanisation of local honey bees and the arrival of *Varroa* mite, Valle, Guzmán-Novoa, Benítez, and Rubio[230] experimented with the Moeller two-queen system. They employed new queens in colony splits, doubling honey yields.

Steve Victors[231] employed cold-adapted Carniolan two-queen colonies built from package bees. The Alaskan beekeeping season is extremely short and it is challenging to build bees to fully exploit honey flows. Victors found that with package bees, headed by cold climate Carniolan stock, two-queen colonies achieved both more rapid colony buildup and doubled honey yields.

While the principle purpose of operating two-queen colonies units has been to vastly improve honey production, their potential for use as powerful pollinators – with minimal supering to facilitate their ready movement – would appear to have been largely overlooked. As we have seen two-queen hives have been adopted for other purposes, notably as means to requeen colonies, as efficient producers of section comb honey, to reduce swarming, to accelerate development of nuclei and – with special measures – to overwinter an additional queen.

Two-queen hives in retrospect

In concluding this overview of the historical development of two-queen hives, we can reflect on the different styles of setups that have evolved over 120 years (Figure 2.23). However the use of standard gear and the adoption of the Moeller-Hogg consolidated brood nest setup (Figure 2.23c) for fixed apiaries and employing variants of the original Farrar two-queen hive for establishing new queens, requeening and strengthening nuclei may help to resolve a range of problems associated with having bees in good condition for honey flows.

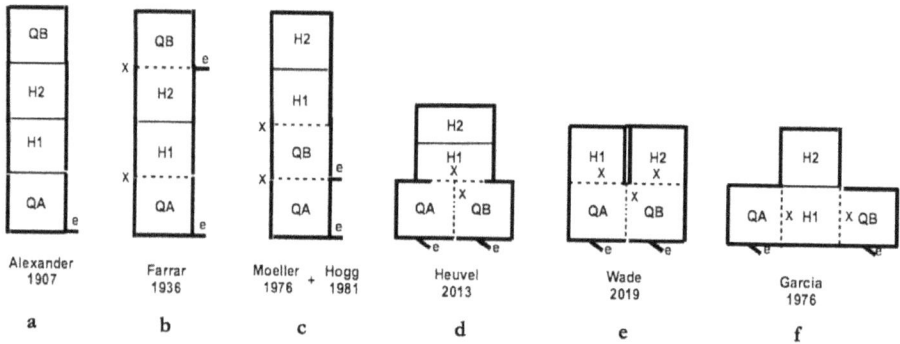

| a | b | c | d | e | f |

Figure 2.23 Historical two-queen hives[232]: e=entrance; x = excluder; CBN = consolidated brood nest; H1, H2 honey supers; QA, QB queens:
(a) Alexander's multiple-queen hive;
(b) Farrar's top-bottom brood nest two-queen hive;
(c) Moeller – Hogg CBN two-queen hive;
(d) Heuvel's horizontal CBN two-queen hive;
(e) Wade's horizontal CBN two-queen hive; and
(f) Garcia's horizontal two-queen hive.

The practices of John Holzberlein and Winston Dunham particularly commend themselves as models for more efficient apiary management. They argued that the average performance of all hives in an apiary is the key determinant of apiary productivity, these days not just the size of the honey harvest but also the income generated from pollination contracts.

Canberra Region Beekeeper members and the club operate both Hogg-Moeller vertical (Figure 2.24) and horizontal (Figure 2.25) consolidated brood nest (CBN) hives. In the next section we will discuss the theory and practice of actually operating two-queen hives.

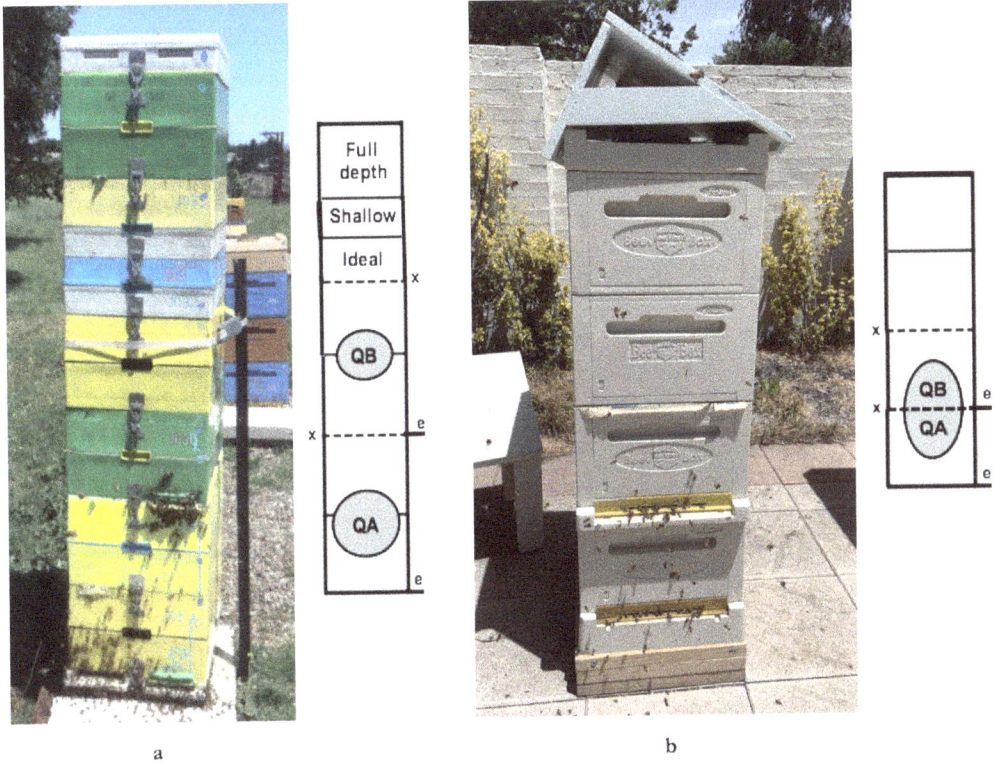

Figure 2.24 Vertical CBN two-queen hives: e = entrance; x = excluder; Q = queen:
(a) Canberra Region Beekeepers' vertical ten-frame Technoset hive December 2018; and
(b) John Robinson's nine-frame Paradise (high density polystyrene) hive November 2019.

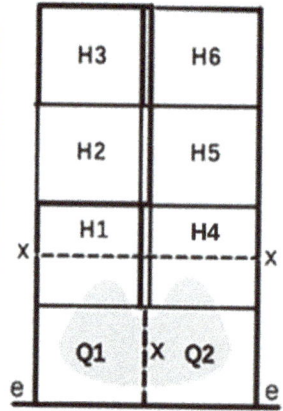

Figure 2.25 Canberra Region Beekeepers' horizontal two-queen hive 1 December 2019:
e = entrance; x = excluder; Q = queen; H = honey super.
Note an oversized Langstroth chamber is divided into two eight-frame compartments
by vertical excluder and supplemented by ideal boxes to provide equivalent
2x12-frame brood chambers, while supers are all eight-frame.

VI — Two-Queen Hive Management

The main problem in establishing two queens in the hive is acceptance of the queens by the bees, just as in single queen operation.

John Hogg

Special measures are required to establish two queens in the one hive. Only then, and with considerable effort, can a two-queen hive be operated successfully.

As we have seen two-queen hives do not set themselves up and when they do they automatically revert to the single-queen condition. So let us make a detailed examination of methods that allow their establishment, the variable broodnest configurations possible, how to actually manage these colonies during the buildup phase and how then to operate them efficiently both during and after the flow.

Most of the principles employed to operate two-queen hives, once established, are much the same as those used in single-queen hive management. However where a single super needs to be added, to facilitate nectar ripening and honey storage in single-queen hive operation, at least two must be added to the two-queen hive. Further, honey must be removed as quickly as possible to warrant the effort of operating colonies with an extra queen.

Essential two-queen hive operational requirements

To start with we can set down how an overwintered single-queen colony can be upscaled to the two-queen condition and how two overwintered single-queen colonies can be united to produce a two-queen hive. From there we can explore the exigencies of management of the just-established two-queen unit.

Two-queen hives are fickle creatures and special care is needed both to ensure that they remain in an optimal condition during the colony building phase and that their two queens are sustained for as long as is needed. There is also an absolute imperative to harvest honey regularly and adjudge when to down-regulate colonies to realise the full potential of the time-limited presence of a second queen.

More skill and effort is required to operate two-queen hives than to manage single-queen hives though their management, too, is very often less than optimal.

Farrar's model for two-queen hive establishment

Traditional two-queen setups employ the Farrar plan pioneered between 1934 and 1946. His plan was to start by installing a new queen on top of a strong hive, that is in a separate and independent nucleus colony perched – like an attic – under the same roof and above a division board or double screen (Figure 2.26). The nucleus is furnished with brood and bees taken from hive below and, of course, a new queen must be supplied.

Three to four weeks later when the fledgling upper colony has its own emerging brood and young workers it is simply united with the parent hive below to form a two-queen hive. To effect the union, the only tools needed – other than a hive tool and smoker – are a queen excluder and a sheet of newspaper. When a double screen is used, rather than a solid division board, no newspaper is required. It is simply a matter of replacing the screen with a queen excluder.

This time-honoured approach for starting up two-queen hives found fuller expression in the 1954 *American Bee Journal* series entitled *How to use two queens for automatic requeening, swarm control and crop increase.*

Contingencies for Farrar's plan included allowing the queenless split to raise its own queen and the opportunity to employ swarm or emergency queen cells – if they were available – in lieu of supplying a mated queen. However Farrar was of the firm view that using nuclei with newly established queens would avoid unnecessary delays in the establishment of two-queen hives.

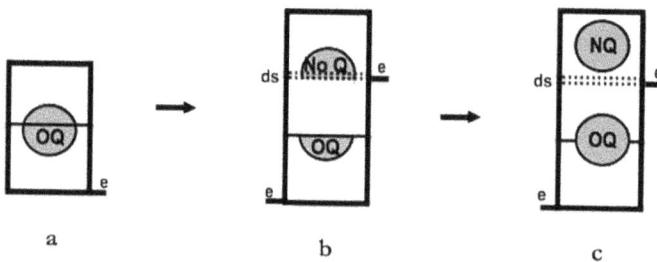

Figure 2.26 Farrar plan for establishment of a second queen in a hive: e = entrance; ds = double screen; OQ = old queen; No Q = no queen: NQ = new queen:
(a) the colony is built to a strong double by reversing the brood boxes and feeding it as needed;
(b) the colony is split with some brood and stores lifted above a nucleus (split) board or double screen where the queen is found and located in the bottom brood chamber; and
(c) a new queen is introduced and established as a prelude to establishing a two-queen hive.

In a routine mid spring inspection of club hives in 2019, I came across a double brood chamber replete with queen cells (Figure 2.27). From all appearances that hive would have swarmed a day or two later but, as it was late in the day and I had many more colonies to inspect, I simply split the hive, supered each unit, stacked the hives and walked away.

Surprisingly not only did the hive abandon swarming but two new queens were found laying well when I inspected the piggybacked hives a little over a month later. The old queen had been superseded and each split had a new queen. The swarm cells that had threatened to result in swarming were repurposed in supplying two exceptionally well raised queens.

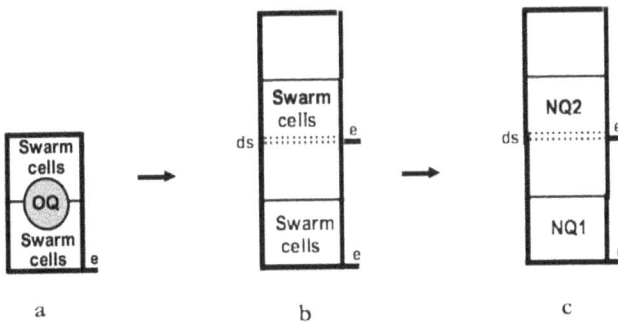

Figure 2.27   Opportunistic swarm control with self double queening: e = entrance; ds = double screen; OQ = old queen; NQ = new queen:
(a) a colony was found in an advanced stage of preparation to swarm;
(b) the colony was split on the vertical pattern and supered to further arrest swarming; and
(c) the new queens had emerged, mated and had each laid out about six frames of brood.

Canberra Region Beekeepers' model for two-queen hive establishment

In our district, the Farrar plan provides insufficient time to build powerful two-queen colonies to take advantage of early flowering Red Box *(Eucalyptus polyanthemos)* that provides a regular calendar-October flow. To capture this flow we adopted an alternative approach to establish our two-queen colonies. Recalling George Wells scheme to prepare his bees at the close of the previous season (there to establish pairs of hives to be ready for use in early spring) we adopted his scheme making up pairs of colonies in autumn to make up two-queen colonies.

In February or March, at the end of the southern hemisphere summer, strong colonies are split: either the old queen is removed and both splits are requeened or the old queen is retained and a new caged queen or cell is introduced to the queenless portion (Figure 2.28). We then follow up to check that both queens are well established and that there are ample stores to optimise their chance of overwintering well. For now, much of Australia remains *Varroa* free, so provided other health checks are made, most well provisioned hives with young queens, make it through to spring.

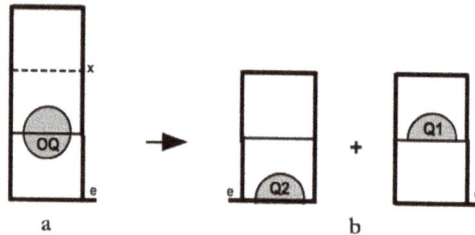

Figure 2.28   End of season hive splitting to form two juxtaposed colonies for overwintering: e = entrance; x = excluder; OQ = old queen; Q1 and Q2 either an old queen and a new queen or two new queens:

(a) strong colonies are selected at the end of the season, with surplus honey supers removed; and

(b) each colony is split, queened and stores added to facilitate overwintering.

This is illustrated for split doubles that were requeened, supered and fed heavily in autumn 2021 (Figure 2.29). Additional support measures included plugging migratory lid vents, filling out the space under migratory covers with insulating material, storing excluders under lids as disease barrier transmission contingency and employing screened bottom boards.[233]

Figure 2.29   Juxtaposed wintered doubles with autumn installed queens and full supers of stores as at 13 August 2021

Then at the earliest opportunity, about three weeks before the spring equinox, this pair of autumn requeened colonies will be reorganised and, once settled, united (Figure 2.30) to from an advanced two-queen colony ready for early Red Box bud break.

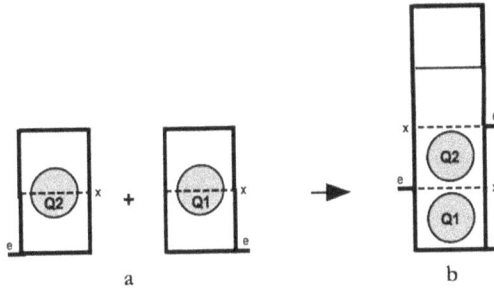

Figure 2.30   Uniting overwintered pair of hives to form an advanced consolidated brood nest two-queen hive: e = entrance; x = excluder; Q1 and Q2 either an old queen and an introduced queen or two autumn-introduced queens:
(a) frames in each colony are sorted so that all unsealed brood is located in bottom brood chambers with excluders employed to locate queens in lower chambers; and
(b) the colonies are united by piggybacking these overwintered doubles using newspaper and an excluder to keep the queens apart.

The process of uniting two colonies to become two-queen hives is best conducted before gear gets too heavy or too difficult to reorganise.

Somewhat later, when queens become commercially available in early October, we establish nucleus colonies using purchased queens as contingency for any two-queen hive failure.

So far we have examined two practical techniques for reliable establishment of two-queen hives. The choice of which scheme to use lies in a clear understanding of the flowering time (phenology) of plants and the climatic factors peculiar to the district.

Hogg's model for two-queen hive establishment

In a radically new approach to two-queen hive establishment Hogg first set out to define the many factors, applying to single-queen as well as multiple-queen hive setups, that influence successful queen introduction. The essential requirements are:

- to make a colony gyne (queen) free before a new queen or queen cell is introduced. This is an absolute prerequisite. The colony must be cleared not only of its reigning queen but also of any additional supersedure queen, virgin queens as well as ripe and developing queen cells. Further it must be free of large numbers of laying workers (gynacoid queens), a condition induced by an extended period of colony queenlessness;

- to time queen introduction to coincide with a nectar flow. This, or feeding bees sugar water to simulate a honey flow, greatly increases the chance of queen acceptance;

- to time queen introduction to avoid periods when bees are ill-tempered. Wait till thunderstorms pass and avoid the other conditions where you are likely to get stung;

- to adopt disruptive techniques that mask the colony defence alarm system such as smoking bees well or spraying them with lavender water;

- to employ queen release delaying tactics, think mailing cage queen candy, to allow the new queen to release pheromones to bees in residence;

- to match the condition of the queen to be introduced to those of the laying queen being replaced, especially critical to the introduction of two queens;

- to use a nucleus colony to requeen where attendant nurse bees and frames of brood and stores promote conditions very similar to those of the parent colony queen; and

- to release any queen or gyne (e.g. a ripe queen cell) into a small colony containing only young nurse bees and emerging bees obtained by shaking older working bees from brood frames.

From this queen acceptance model Hogg was able to devise several schemes where two queens, kept apart by a queen excluder, could be reliably established. His elegant consolidated brood nest two-queen system set his scheme apart from most of those formerly practiced:

The consolidated brood nest hive has two brood chambers separated by a single queen excluder with a queen in each chamber. There are two entrances which face in opposite directions (lower front and upper rear).

Brood is continuous throughout the two chambers. The brood nest is indistinguishable from that of a single queen occupying the two brood chambers. Hence the term consolidated brood nest (CBN).

All bees have access to both queens and their brood, and to common supers above for storage of surplus honey. All of the nurse bees presumably participate in brood rearing and in the distribution of queen substances throughout the entire consolidated brood nest.

Hogg defined three disparate approaches to two-queen hive establishment outlined in detail in the May and June 1983 issues of the *American Bee Journal*[34].

(a)   Simultaneous direct introduction of two laying queens, virgin queens or queen cells

This technique requires both queens be in a very similar condition (e.g. mated caged queens of the same provenance and age) and that each brood nest be closely matched for condition. Further bees in each unit must be trained to use their own entrance so that, once united, the bees do not find the need to use each other's brood chamber entrance (Figure 2.31).

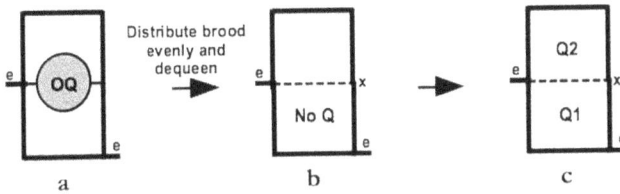

Figure 2.31   Direct double queening starting with a double brood chamber with bees conditioned to employ two entrances: e = entrance; x = excluder; OQ = old queen; No Q = no queen; Q1 and Q2 = new queens:
(a) a strong double colony with a double brood chamber is identified and an upper entrance is provided to condition bees to fly to discrete entrances;
(b) the brood is evenly distributed, dequeened and an excluder is inserted; and
(c) two new queens are introduced simultaneously.

Since both brood chambers are in the same condition and bees all come from the same colony, no precautions are needed to unite the two queens.

(b)   Uniting queenright colonies

This method is probably the least risky and most assured of success, one we have used extensively. Each chamber is first acclimatised for flight entrance condition where the units are first stacked (separated by a nucleus board or a double screen) or placed alongside each other. In this example (Figure 2.32) a strong colony is split in mid spring and the new and old queen (or two new queens) are established in separate stand alone colonies. Once well established the single chamber colonies are simply papered together always locating any older queen below. Single colonies (or nucs) from an out-apiary can be entrance trained for a few days and then directly united avoiding the first step.

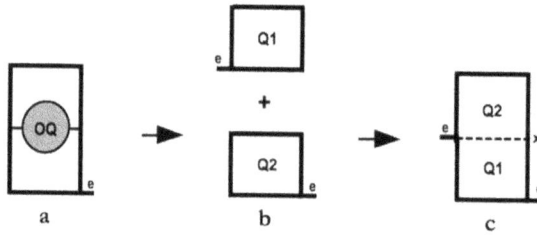

Figure 2.32 Uniting two queenright colonies to form a two-queen brood nest: e = entrance; x = excluder; OQ = old queen, Q1, Q2 old and new queens or new queens of the same provenance:

(a) a strong overwintered colony is built;

(b) the colony is split, brood, bees and stores are evenly distributed, the old queen is removed and queens are introduced and allowed to establish well; and

(c) the split colonies are united using an excluder and newspaper.

A simple variant (Figure 2.33) was used by Hogg to set up consolidated brood nest hives.

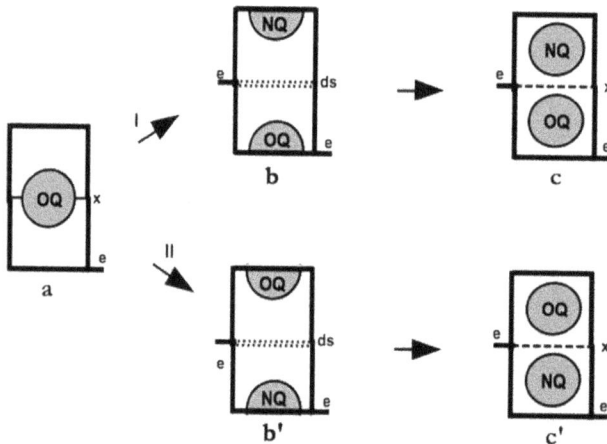

Figure 2.33 Hogg's two step process for uniting queenright colonies to produce a two-queen brood nest: e = entrance; ds = double screen; x = excluder; OQ = old queen; NQ = new queen:

(a) a strong double brood chamber hive with an overwintered old queen is identified;

(b and b') the colony is split by the simple insertion of a double screen with no attempt to find the old queen. This creates Demaree like conditions conducive to the raising of supersedure cells: a caged queen is introduced to the unit that commenced building queen cells or that does not have eggs several days later; and

(c and c') after a little over three weeks, when the new queen has her own emerging brood, the double screen is replaced by a queen excluder to form a two queen hive.

Pathway I

In this event the upper unit with some sealed brood will benefit from the rising warmth of the unit below and from emerging worker bees, both conducive to raising a new queen if one is not supplied. Progress of the introduction of the upper queen can be monitored by simply lifting the hive lid to check for establishment of the new queen and the subsequent presence of eggs and young larvae. In the interim, and if the unit is very strong, there is some risk of the lower unit becoming too crowded and swarming, a condition that can be alleviated by under-supering the double screen. This however, defeats the purpose of direct establishment of a double chamber consolidated brood nest (CBN).

Pathway II

In this setup the lower super will take some time to establish so will be at reduced risk of swarming. However the upper unit can be supered without upsetting the CBN brood nest arrangement.

(c)   Direct introduction of a queen to a queenright colony

At first sight this approach would seem to defy the fundamental one colony – one queen principle. Try to introduce a new queen to an already queenright colony and, in the normal course of events, that queen will be lost. However, if an upper chamber is conditioned to be in a semi-autonomous state, then a queen can be introduced to the queen free zone.

To precondition the colony bees must first be trained to use either an upper or a lower hive entrance ideally both having been open for some weeks. The brood frames are then sorted placing sealed brood only in the upper brood chamber. Unsealed brood, all located in the lower brood nest, will attract nurse bees down through the excluder, but only as needed, while nurse bees tending the lower queen brood nest will not be drawn upwards.

With these preparations, a caged queen can be introduced to the upper unit. Newly emergent bees above the excluder will be receptive to the new queen – and migrate down through the excluder if surplus to needs – while field bees will continue to use their respective entrances. For a while the units above and below the excluder, though not isolated, will operate more or less independently (Figure 2.34).

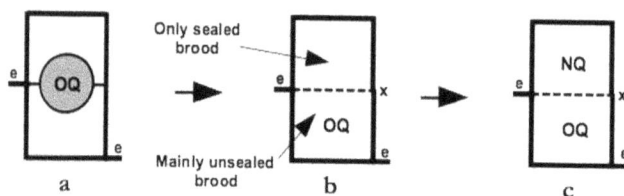

Figure 2.34 Direct queening of a conditioned queenright colony to effect two-queen establishment: e = entrance; x = excluder, OQ = old queen, NQ = new queen: (a) a strong single-queen hive occupying a double brood chamber is reversed regularly to distribute brood evenly with bees trained to use either the upper or lower entrance; (b) frames are sorted so that only sealed brood is present in the upper chamber, an excluder is inserted and bees are shaken down or the old queen is found and located below the excluder; and (c) a new queen is immediately introduced to the top brood chamber to effect direct establishment of a two-queen colony.

This technique is reminiscent of E.W. Alexander's 1907 discovery that, with exceptional care and without an excluder, a conditioned queen can be introduced to an already queenright colony.

Building two-queen colonies

With this understanding of practical setup of two-queen hives, we can turn to requirements for their ongoing management. Two-queen tower hives can be operated in two distinctive ways.

(a)   Historical top-bottom brood nest two-queen hive operation

In the traditional model of two-queen hive operation, one queen is confined to a lower brood box – sometimes constituted as a double chamber – above which are placed honey supers. Towards the top of this hive there is an additional and entirely separate chamber containing the second queen.

The overall setup is one of top-and-bottom brood chambers with honey supers sandwiched between. Additional honey supers are inserted as needed and then rotated out (Figure 2.35, *Pathway I*). The principles are the same as those for single-queen operation except in that more supers are required and honey can only be harvested by temporarily offsetting the upper brood chamber. With strong honey flows, the operation of such mega colonies becomes a logistics nightmare. The alternative scheme adopted by Floyd Moeller and John Hogg was to consolidate the brood nest by moving the upper brood nest down to enable colony to operated in a manner identical to that for operation of a single-queen hive (Figure 2.35, *Pathway II*). The more ready access to honey supers greatly simplifies honey harvesting and – in very advanced two-queen hive operation – has found renewed application in section comb honey production[235].

The traditional scheme (Figure 2.35, *Pathway I*) has found other more specialist applications. These include temporary establishment of new queens and nucleus colony strengthening and supplying spare queens, brood and nuclei to rectify problems encountered with struggling hives. As Dunham concluded:

This plan is very versatile and allows also for preflow conditioning, making increase, and testing queens.

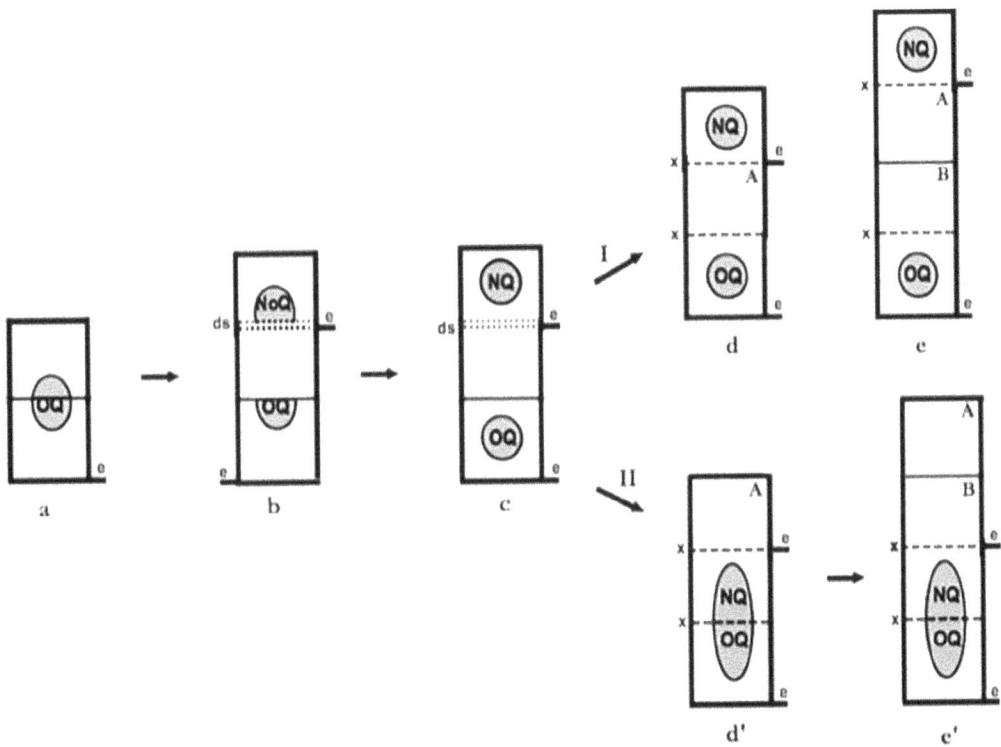

Figure 2.35 Traditional Farrar two-queen scheme and alternative Moeller-Hogg consolidated brood nest operations: e = entrance; ds = double screen; x = excluder; OQ = old queen; NQ = new queen:
(a) an overwintered double is reversed regularly to stimulate brood rearing;
(b) a doubled screen is inserted, over an added super of drawn combs, splitting the brood nest with bees shaken down to locate the old queen below;
(c) a new queen (or cell) is introduced and given 3-4 weeks to become well-established;

Pathway I

(d) the double screen is replaced by a queen excluder to form the classic top-bottom Farrar-style two-queen colony; and

(e) supers are progressively sandwitched between the brood nests – though the top brood chamber may also be supered.

Pathway II

(d') the double screen is replaced by a queen excluder but the supers are reorganised to form a consolidated brood nest two-queen hive; and

(e') the colony is progressively supered providing more space and undersupering returned extracted combs as for single-queen hive operation.

One barrier to Farrar's scheme was the seemingly necessary requirement to acquire a crane driver's licence (Figure 2.36). This was needed to remove honey literally pouring into his tower hives. In my view this scheme is no longer the method of choice for operating two-queen hives. This said, Farrar was far and away ahead of his time proffering a number of helpful tips[236] for operating two-queen hives noting:

> *After the flow begins, supers are given to both units but the supers placed between the two excluders must not be allowed to fill up or you will have in effect two separate colonies.*

Figure 2.36   Farrar's two-queen tower hives employing shallow square supers throughout

(b)   Consolidated brood nest two-queen operation

Tower hive operation is a more or less a given for any two-queen hive operation, but is greatly simplified where the brood nests are consolidated and where supering for the flow can be conducted brood-free above an excluder. While we have shown that a horizontal consolidated two-queen brood nest configuration overcomes the problem of operating overly tall hives, the need to employ non-standard oversized divided brood chamber mitigates against its wide use.

Two-queen hive brood nest establishment can be staged to commence in autumn where split colonies are queened, overwintered as pairs of hives and united to form two-queen colonies at the first opportunity in spring. The alternative is to delay all operations until spring when two-queens hives should be established from a strong overwintered hive by one means or another and as early as possible. The tradeoff is the need to use two overwintered colonies versus much later establishment of two-queen hives from single-queen hives. With a two-queen brood nest established a generalised scheme for first building bees and then operating them through the flow can be formulated (Figure 2.37).

Firstly the two-queen unit must be built in time for the earliest major flow, that is allowing both queens to lay together for seven to eight weeks. The fully integrated broodnest will require sufficient space for two queens laying at full tilt and will require double the amount of pollen and stores to support a much larger developing workforce: two children eat more than one. The dynamics of a doubled brood nest containing two queens is entirely different to that of a single-queen hive. For example a starting single-queen colony must be very strong if, in its split state, each of two queens is to have enough bees for each of them to build quickly.

Then as brood rearing tapers off and the honey flow starts, the relatively high efficiency of a giant population must be factored in to supering and honey harvesting. While the presence of two young and well matched queens largely makes two-queen hives swarm proof, the countervailing *bees need space* edict of Don Peer means that two-queen hive honey flows must given full attention and room to operate. Bees can only cure incoming nectar and store honey when supers are added or returned to the hive as extracted sticky combs.

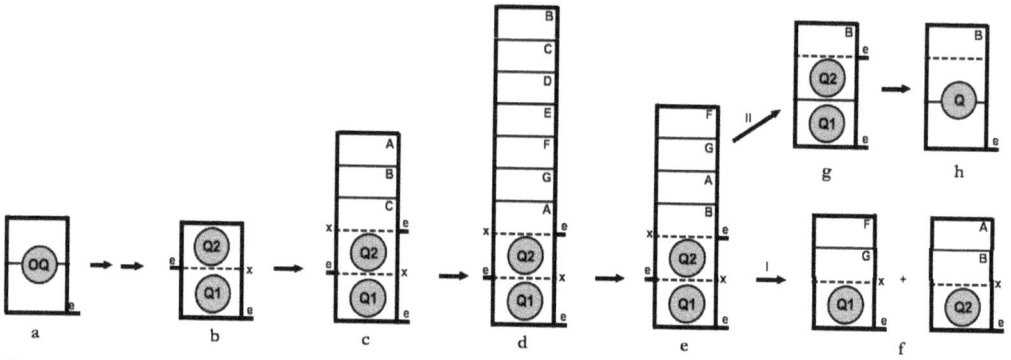

Figure 2.37 Generalised scheme for operating a CBN two-queen hive: e = entrance; x = excluder; OQ = old queen; Q1, Q2 queens:

(a) a starting strong single-queen colony is identified;

(b) colony is established as a two-queen hive – by any of the various means described;

(c) the colony is supered as needed to accommodate an expanding population;

(d) the colony is heavily supered as the main honey flow commences: supers that have fully capped honey (supers such as A) are removed, are extracted and are returned undersupering the stack while supers (here designated as supers B and C) that are nearly filled are placed on top of the stack; and

(e) as the flow passes its peak filled supers at the top of the stack are either extracted and returned (super B) or progressively removed (supers C, D and E); and

Pathway I

(f) the colony is split as separate colonies, requeened if possible as well as fed as needed; or

Pathway II

(g and h) the queen excluder is removed and left with the equivalent of a full depth box of stores allowing one queen to supersede the other.

Note: All excluders should be removed and can be stored under hive lids for wintering.

The overall working of this scheme, though seemingly complex, follows closely that for single-queen hive management particularly in respect of supering for the flow and harvesting honey, albeit the two-queen hive being operated as a colony on steroids.

On a more practical note I have found brood boxes in two-queen hives extremely difficult to inspect during a honey flow and rarely attempt to do so. I have also found that attention must be given to removing surplus gear and reducing the colony to a single-queen condition once the party – the honey flow – is well in progress and especially when the flow is nearly at an end.

Here John Holzberlein's observation *that the answers are not all apparent at once, nor will any set of rules work all the time* rings especially true. It is neither possible to have a forward plan to cover all seasonal contingencies nor is it possible to guarantee that, especially in a drought year, two-queen colonies will not revert to the single-queen condition and that hives may require only minimal supering.

Operating two-queen hives always proves challenging though I have concluded that it can only be done successfully if the normal practices of regular requeening, establishing good disease barriers, ensuring there are adequate stores – particularly in the buildup phase – and providing adequate space for the queen to lay and for bees to process and store honey are paid close attention to.

To conclude I would like to cite examples of successful operation of two-queen hives. There have of course been failures, some on the bees part and not quite all of my own making. I put most colony failure down to use of queens more than a year old. There I have found that use of an older queen often results in loss of one queen especially whenever there is an interruption to the flow.

Example 1

The hive (Figure 2.38), along with a horizontal two-queen hive and a doubled hive, produced the bulk of the club honey harvest – amongst a sea of single-queen hives – in the Canberra Region Beekeepers' bumper summer season of 2018-2019.

Figure 2.38 Ten-frame fully supered two-queen hive setup at Canberra Region Beekeepers' apiary at Jerrabomberra Wetlands, 1 December 2018: e = entrance; x = excluder; Q = queen; FD = full depth super; ID = ideal super; MA = Manley super:

(a) front of hive;

(b) back of same hive rear; and

(c) schematic showing queens housed in single brood chambers and mixed depth honey supers.

Example 2 This hive (Figure 2.39) is representative of Beeblebox Apiary two-queen hives that were very successfully operated over the same 2018-2019 season employing eight frame gear. In the subsequent drought seasons of 2019-2020 and 2020-2021 two-queen hives, those that did not revert to single-queen colonies, produced only a modest crop that was at least comparable to similarly operated single-queen hives.

The experience of mixed performance of two-queen hives in poor seasons, when the long term forecast is for very dry conditions, suggests that the effort and cost of running two-queen hives may not then be warranted. Ever the optimist, with normal to above average rainfall forecast for the season ahead and with eucalypts in the heaviest bud since the 1980s, it will be just too tempting not to put hives with two queens to work.

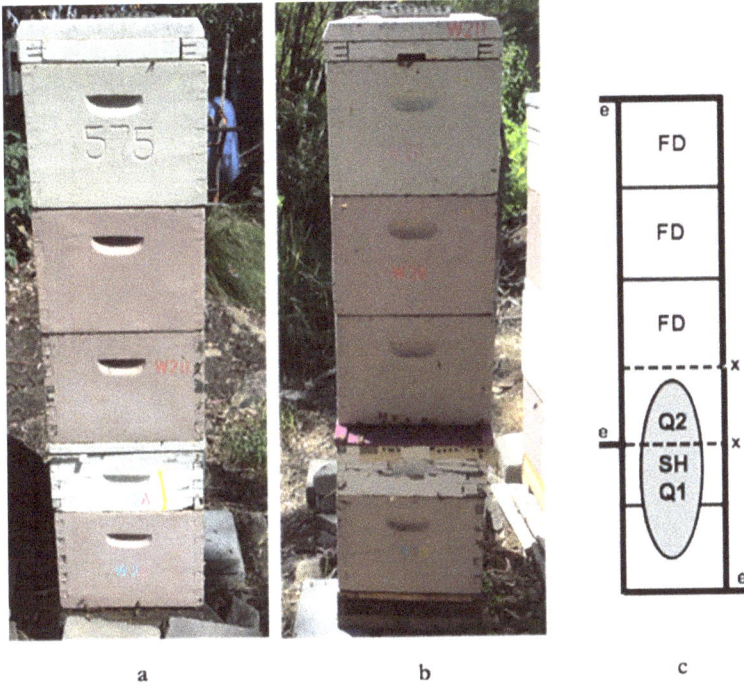

Figure 2.39   Eight-frame fully supered two-queen hive setup in author's backyard Beeblebox Apiary, 17 December 2018: e = entrance – note top rear entrance; x = excluder; Q = queen; FD = full depth super; SH = shallow super:

(a) front of CBN eight frame hive;

(b) back of same hive; and

(c) eight-frame schematic for eight frame operation.

. . . .

Acknowledgements

I would like to thank Liz Milla at the CSIRO Black Mountain Herbarium, Anne-Marie Slattery, Josie Braddick and William Hastie at the CSIRO Black Mountain Library and Michael Herlihy from the Australian National Library for their assistance in locating publications. I would also acknowledge Dannielle Harden, Frank Derwent and John Robinson from the Canberra Region Beekeepers' *Two-Queen Hive Special Interest Group* for their contributions. Finally I would like to thank Dave Flanagan for his contribution on tripled hive beekeeping on the Paroo River.

1   Wells. G. (1894). *Guide Book pamphlet on the two-queen system of bee keeping.* G.F. Gay. Snodland Steam Printing Works, Malling Road, 15pp.

2   British Bee-Keepers' Association Quarterly Converzatione (October 18, 1894). *British Bee Journal, Bee-Keepers' Record and Adviser* **22(643):**411-413.

3   Demuth, G.S. (1921). Swarm control. *Farmers Bulletin* **1198:**1-45. https://ia601401. us.archive.org/32/items/CAT87202908/farmbul1198.pdf

4   Bee (1904). Wells hive and system. *West Gippsland Gazette,* 17 May 1904, p.4. https://trove. nla.gov.au/newspaper/article/68708484?searchTerm=wells%20system&searchLimits=#

5   British Bee-Keepers' Association Quarterly Converzatione (April 7, 1892). Report of meeting of 31 March 1892. *British Bee Journal, Bee-Keepers' Record and Adviser* **20(511):**132-133.

6   British Bee-Keepers' Association Quarterly Converzatione (April 6, 1893). *British Bee Journal, Bee-Keepers' Record and Adviser* **21(563):**126, 132-134. Editorial notices &c. (November 12, 1896). British Bee-Keepers' Association Conversazione. *British Bee Journal, Bee-Keepers' Record and Adviser* **24(752):**451-453.

7   Editors British Bee Journal (May 21, 1891). When doctors and professors differ! A young queen hatching in a hive above queen-excluder placed on the old brood chamber. *British Bee Journal, Bee-Keepers' Record and Adviser* **19(465):**239-240.

8   Tinker, G.L. (March 10, 1892). Queens fertilised in full colonies with a laying queen. *British Bee Journal, Bee-Keepers' Record and Adviser* **20(507):**95-96.

9   Laurence, H. (April 5, 1894). A lady's bee experience. *British Bee Journal, Bee-Keepers' Record and Adviser* **22(615):**133-134.

10   As communicated by George Wells great great granddaughter, Sarah Austin (pers. comm.).

11   Northumberland and Durham Bee-Keepers' Association (22 February 1894). The Wells System: Lecture by Mr Wells. *British Bee Journal, Bee-Keepers' Record and Adviser* **22(609):**79.

12   Northumberland and Durham Bee-Keepers' Association (March 8, 1894). *British Bee Journal, Bee-Keepers' Record and Adviser* **22(611):**100.

13   Editors British Bee Journal (April 20, 1893). The double-queen system: A visit to Mr Wells' apiary. *British Bee Journal, Bee-Keepers' Record and Adviser* **21(565):**151-153.

14   Mr Wells's apiary (July 6, 1899). Homes of the honey bee: The apiaries of our readers. *British Bee Journal, Bee-Keepers' Record and Adviser* **27(889):**262-263.

15   Martin, W. (October 4, 1894). The Wells hive: From a Cottager's point of view. *British Bee Journal, Bee-Keepers' Record and Adviser* **22(641):**396.

16   Nicholls, A. (August 30, 1894). My first experience with a Wells hive. *British Bee Journal, Bee-Keepers' Record and Adviser* **22(636):**346.

17   The Northumberland and Durham Bee-Keepers' Association (February 8, 1894). The Wells System: A lecture by Mr Wells. *British Bee Journal, Bee-Keepers' Record and Adviser* **22(607):**52.

18   Kidd, J.N. (1921). Relative to the two queen system. *Bee World* **3(7):**192-193. https://doi. org/10.1080/0005772X.1921.11095216

19   Wells Hives. No.1. Meadows' Wells hive. (March 16, 1893). *British Bee Journal, Bee-Keepers' Record and Adviser* **21(560):**103. Wells Hives No.2. Blow's Wells hive. (March 23, 1893). *British Bee Journal, Bee-Keepers' Record and Adviser* **21(561):**114.

Wells hives. No.3. Neighbour's Wells hive. (March 30, 1893). *British Bee Journal, Bee-Keepers' Record and Adviser* **21(562):**125.

Wells Hives No.3. Howard's Wells hive. (April 6, 1893). *British Bee Journal, Bee-Keepers' Record and Adviser* **21(563):**135.

Wells Hives. No.4. A.W. Harrison's Wells hive. (April 6, 1893) *British Bee Journal, Bee-Keepers' Record and Adviser* **21(563):**135.

Wells Hives. No.6. Redshaw's Wells hive. (June 22, 1893). *British Bee Journal, Bee-Keepers' Record and Adviser* **21(574):**243.

E.C. Walton Wells hive (June 7, 1894). *British Bee Journal and Bee-Keepers' Adviser* **22(624):**210.

20   Drallop (Taylor, E.H.). The Ford-Wells hive. (February 1, 1906). *British Bee Journal and Bee-Keepers' Adviser* **34(1232):**49.

21   Mr Rymer's Wells Apiary (March 17, 1898). Homes of the honey bee: The apiaries of our readers. *British Bee Journal and Bee-Keepers' Adviser* **26(821):**104-106.

Mr Hall's Apiary (February 15, 1906). Homes of the honey bee: The apiaries of our readers. *British Bee Journal and Bee-Keepers' Adviser* **34(1234):**64-66.

22   Woodley, W. (November 17, 1892). Notes by the way. *British Bee Journal, Bee-Keepers' Record and Adviser* **20(543):**447 refers to George Wells' trial of doubled v. single-queen hives at Wells, G. (November 3, 1892) loc. cit.

Wells, G. (November 24, 1892). The Wells System. *British Bee Journal, Bee-Keepers' Record and Adviser* **20(544):**459-460.

Wells, G. (May 18, 1893). Comparing the double and single-queen systems. *British Bee Journal, Bee-Keepers' Record and Adviser* **21(569):**195-196.

23   E.B. (February 23, 1893). Results of the Wells System: The reason why. *British Bee Journal, Bee-Keepers' Record and Adviser* **21(557):**77.

24   Editors British Bee Journal (April 20, 1893). The double queen system: A visit to Mr Wells' apiary. *British Bee Journal, Bee-Keepers' Record and Adviser* **21(565):**151-153.

25   Ward, T.F. (February 21, 1895). The Wells System, and Mr Wells' reply. *British Bee Journal and Bee-Keepers' Adviser* **23(661):**78.

26   Wells, G. (February 28, 1895). Mr Wells and his critics. *British Bee Journal and Bee-Keepers' Adviser* **23(662):**89.

27   Wells, G. (February 14, 1895). The double-queen system: Mr Wells' reply to his critics. *British Bee Journal and Bee-Keepers' Adviser* **23(660):**64-65.

28   The Village Blacksmith (May 5 1892). Two queens in one hive. *British Bee Journal, Bee-Keepers' Record and Adviser* **20(515):**176.

Wood, A.J.H. (May 5 1892). Working with two queens in one hive. *British Bee Journal, Bee-Keepers' Record and Adviser* **20(515):**176.

Jones, B.E. (May 17, 1894). The two-queen system: The season in the Fielded District, Lancashire. *British Bee Journal, Bee-Keepers' Record and Adviser* **22(621):**194-195.

Wells, G. (May 31, 1894). The Wells System. *British Bee Journal, Bee-Keepers' Record and Adviser* **22(623):**212.

Rothery (May 31, 1894). Early honey in Yorkshire: Double-queened hives. *British Bee Journal and Bee-Keepers' Adviser* **22(623):**213-214.

Wood, A.J.H. (June 7, 1894). Single v. double queened hives. *British Bee Journal and Bee-Keepers' Adviser* **22(624):**225.

Wells, G. (June 7, 1894). Double-queened hives: Who originated the idea? *British Bee Journal and Bee-Keepers' Adviser* **22(624):**225-226.

Hawke-eye (June 7, 1894). Early Yorkshire honey and the claimant of the Wells System. *British Bee Journal and Bee-Keepers' Adviser* **22:(624):**226.

29   Simmins, S. (April 15, 1908). Plurality of queens: Points in the Wells System explained. *Gleanings in Bee Culture* **36(8):**506-507. https://babel.hathitrust.org/cgi/pt?id=uc1. b3458202&view=1up&seq=518

30   Simmins, S. (1914). A modern bee-farm and its economic management; showing how bees may be cultivated as a means of livelihood; as a health-giving pursuit; and as a source of recreation to the busy man, pp.463-464. Hepworth and Company, Limited, Tunbridge Wells, 165, Queen Victoria St., London, E C. https://ia800202.us.archive.org/3/items/cu31924003428038/cu31924003428038.pdf

31   Stroud, J.W. (1884). The honey bee (*Apis mellifica*): Its natural history and management from a South African point of view. *Transactions of the Eastern Province Naturalists' Society*, **No.1:**5-66. S2A3 Biographical Database of Southern African Science https://www.s2a3.org.za/bio/Biograph_final.php?serial=2755
Stroud, J.W. (1884). The honey bee (*Apis mellifica*): Its natural history and management. *The Transactions of the Eastern Province (South Africa) Naturalists' Society* Part I:5-66.— By J.W. Stroud, M.D., Port Elizabeth, Cape Colony.
Reviews (May 1 1885). The honey bee by Dr Stroud. *British Bee Journal and Bee-Keepers' Adviser* **13(179):**156.
Stroud, J.W. (September 15, 1886). Honey-bees in the Cape Colony. *British Bee Journal and Bee-Keepers' Adviser* **14(178):**299-300.

32   Stroud, J.W. (April 15, 1885). Foreign: South Africa. *British Bee Journal and Bee-Keepers' Adviser* **13(168):**133.
Walker, G. (May 1, 1885). The honey bee. *British Bee Journal and Bee-Keepers' Adviser* **13(169):**156-157.
Stroud, J.W. (July 1, 1885). South African bees. *British Bee Journal and Bee-Keepers' Adviser* **13(173):**220-221.
Foreign (March 25, 1886). South Africa. *British Bee Journal and Bee-Keepers' Adviser* **14(196):**125-126.
Adams, J.E. (March 25, 1886). African bees. *British Bee Journal and Bee-Keepers' Adviser* **14(196):**129.
Advertisement for South African queen bees. (April 1, 1886). *British Bee Journal and Bee-Keepers' Adviser* **14(197):**146.
British Bee-Keepers' Association second Quarterly Converzatione (May 6, 1886). *British Bee Journal and Bee-Keepers' Adviser* **14(202):**196-198.
Walker, G. (June 24, 1886). African queen. *British Bee Journal and Bee-Keepers' Adviser* **14(203):**282.
Useful hints (July 8, 1886). The various races of bees. *British Bee Journal and Bee-Keepers' Adviser* **14(211):**304.
Simmins, S. (July 15, 1886). Another South African queen. *British Bee Journal and Bee-Keepers' Adviser* **14(212):**316.
Broome, H.A. (August 12, 1886). Notices to correspondents & inquirers. *British Bee Journal and Bee-Keepers' Adviser* **14(216):**379.

33   Heddon, J. (1885). *Success in bee-culture as practiced and advised by James Heddon*, 128pp. Dowagiac, Mich., Times print, p.78. https://babel.hathitrust.org/cgi/pt?id=loc.ark:/13960/t8x931940&view=1up&seq=86

34   Simmins, S. (1893). *A modern bee-farm and its economic management:* Showing how bees may be cultivated as a means of livelihood, as a health-giving pursuit, and as a source of recreation to the busy man. Chapter XXV, Working two queens in one hive. The Wells' System, pp.225-237. London. Woodford Fawcett & Co., 112, Fleet Street, E.G. https://ia800205.us.archive.org/1/items/cu31924003200726/cu31924003200726.pdf
Simmins, S. (1887). *A modern bee-farm and its economic management.* London, T. Pettit &

Co, 22 and 23, Frith Street, Shaftesbury Avenue, W., 227pp. https://ia800206.us.archive. org/24/items/cu31924003200742/cu31924003200742.pdf

35   Neighbour, A. (1878). *The apiary, or, bees, beehives and bee-culture*: Being a familiar account of the habits of bees and the most improved methods of management, p.15. Third edition, Kent and Co., Paternoster Row; Geo. Neighbour and Sons, 149 Regent Street and 127 High Holborn, London. http://the-eye.eu/public//Books/Survival_Guide/Beekeeping/the_ apiary_1878.pdf

36   Dadant, C. (June 17, 1886). Gleanings in *Bulletin d' Apiculture de la Suisse Romande*. *British Bee Journal and Bee-Keepers' Adviser* **14(208):**269.

37   Wells, G. (18 May, 1893) loc. cit.

38   Wells, G. (November 3, 1892). Two queens in a hive: The Wells System. *British Bee Journal and Bee-Keepers' Adviser* **20(542):**438-439.
Wells, G. (December 7, 1893). Mr Wells' report for 1893. *British Bee Journal and Bee-Keepers' Adviser* **21(598):**485-487.
Wells, G. (December 13, 1894). Double v. single-queened hives: Mr Wells' report of season '94. *British Bee Journal and Bee-Keepers' Adviser* **22(651):**493-495.
Wells. G. (December 19, 1895). Mr Wells' report for 1895. *British Bee Journal and Bee-Keepers' Adviser* **23(604):**515-516.
Wells, G. (November 12, 1896). The double-queen system: Mr G. Wells's report for 1896. *British Bee Journal and Bee-Keepers' Adviser* **24(751):**454-455.
Wells, G. (March 31, 1898). Mr G. Wells' report for 1897. *British Bee Journal and Bee-Keepers' Adviser* **26(823):**125-126.
Wells, G. (April 13, 1899). My bee-doings for '98: Mr George Wells' annual report. *British Bee Journal and Bee-Keepers' Adviser* **27(877):**144, 146.
Wells, G. (March 29, 1900). Mr G. Wells's report for 1899. *British Bee Journal and Bee-Keepers' Adviser* **28(927):**127-128.
Wells, G. (December 27, 1900). Mr Geo Wells's annual report. *British Bee Journal and Bee-Keepers' Adviser* **28(966):**508.

39   Wells, G. (May 18, 1893) loc. cit.

40   Wade, A. (2019). Establishing two queens instead of one queen in a honey bee colony Part I: Principles of introducing and running two-queen colonies. *The Australasian Beekeeper* **120(8):**18-21.
Wade, A. (2019). Establishing two queens instead of one queen in a honey bee colony Part II: Setting up two-queen colonies. *The Australasian Beekeeper* **120(9):**16-21.
Wade, A. (2019). Establishing two queens instead of one queen in a honey bee colony Part III: Operating two-queen colonies. *The Australasian Beekeeper* **120(10):**22-25.

41   Farrar, C.L. (1968). Productive management of honey bee colonies. *American Bee Journal* **108(3):**95-97, 141-143.
Farrar, C.L. (1968). Productive management of honey bee colonies. *Apiacta* **4:**22-28. http:// www.fiitea.org/cgi-bin/index.cgi?sid=&zone=cms&action=search&categ_id=55&search_ ordine=descriere

42   Ferris, A.K. (1906). The hive adapted to the two-queen system: Reasons why ten-frame hive is unsuitable. *Gleanings in Bee Culture* **34(9):**586-587. https://babel.hathitrust.org/ cgi/pt?id=uc1.a0003415858&view=1up&seq=346 https://ia800902.us.archive.org/3/items/ gleaningsinbeecu34medi/gleaningsinbeecu34medi.p df
Ferris, A.K. (1906). Comb honey by the two-queen system: Strong colonies for comb and weak ones for extracted honey: The advantage of a dual in each hive. *Gleanings in Bee Culture* **34(12):**803-804. https://babel.hathitrust.org/cgi/pt?id=uc1. a0003415858&view=1up&seq=489
Ferris, A.K. (1906). The Ferris system of producing comb honey and swarm control: A

cheaper comb-honey device. *Gleanings in Bee Culture* **34(18):**1184. https://babel.hathitrust.org/cgi/pt?id=umn.31951d00953179c&view=1up&seq=698

Lawrence, S.T. (September 1926). A two-queen system especially adapted to the sweet-clover region. *Gleanings in Bee Culture* **54(9):**583-586. https://archive.org/details/sim_gleanings-in-bee-culture_1926-09_54_9/page/582/mode/2up

43   Sladen, F.W.L. (1920). The Sladen two-queen system. *American Bee Journal* **60(3):**84-86. https://ia802702.us.archive.org/19/items/americanbeejourn6061hami/americanbeejourn6061hami.pdf

Sladen, F.W.L. (1921). Wintering two queens in one hive. *American Bee Journal* **60(9):**348-349. https://ia802702.us.archive.org/19/items/americanbeejourn6061hami/americanbeejourn6061hami.pdf

44   Fowls, I. (1918). The best from others: Intensive beekeeping: Trial of an intensive system of beekeeping. *Gleanings in Bee Culture* **46:**742. https://ia802708.us.archive.org/27/items/gleaningsinbeecu46medi/gleaningsinbeecu46medi.pdf [This article reports on an article by F.W.L. Sladen, C.E.F., Ottawa that appeared in the October 1918 edition of the *Canadian Horticulturist and Beekeeper* **26(10).**]

Sladen, F.W.L. (July 3, 1919). Combs from other hives: Trial of a system of keeping two queens in a hive. The *British Bee Journal and Bee-Keepers' Adviser* **47(1932):**281-282. Reprinted from Sladen, F.W.L. (February 1919). *Agricultural Gazette of Canada* **6(2):**130-134.

45   Sladen, F.W. (1919). Intensive Beekeeping. *American Bee Journal* **59(4):**118-119. https://ia800907.us.archive.org/21/items/americanbeejourn5859hami/americanbeejourn5859hami.pdf

46   Masheath bee hive advertisements (October 23 and 30, 1919). *British Bee Journal and Bee-Keepers' Adviser* **47(1948-1949):**476, 488.

47   Atkinson, M. (June 10, 1920). Queens and tiered chamber bee spaces. *British Bee Journal and Bee-Keepers' Adviser* **48(1981):**280-282.

48   Atkinson, M. (October 27, 1921). Dual-queen work: The double-six system. *British Bee Journal and Bee-Keepers' Adviser* **49(2053):**501-502.

Atkinson, M. (November 17, 1921). Dual-queen work: The double-six system. *British Bee Journal and Bee-Keepers' Adviser* **49(2056):**537.

49   Ellis, J.M. (October 13, 1921). Notes from Gretna Green. *British Bee Journal and Bee-Keepers' Adviser* **49(2051):**477.

50   Atkinson., M. (October 27, 1921) loc. cit.

51   Atkinson, M. (January 5, 1922). Advertisement for Atkinson's 1920 Cambridge lecture notes. *British Bee Journal and Bee-Keepers' Adviser* **50(2063):**49.

52   Ellis, J.M. (November 26, 1908). Management at the heather: Wanted better methods and results. *British Bee Journal and Bee-Keepers' Adviser* **36(1379):**475-476. https://ia800202.us.archive.org/10/items/britishbeejourna1908lond/britishbeejourna1908lond.pdf

Medicus (December 17, 1908). Management at the heather. *British Bee Journal and Bee-Keepers' Adviser* **36(1382):**504, 506. https://ia800202.us.archive.org/10/items/britishbeejourna1908lond/britishbeejourna1908lond.pdf

Ellis, J.M. (December 31, 1908). Management at the heather. *British Bee Journal and Bee-Keepers' Adviser* **36(1384):**522-523. https://ia800202.us.archive.org/10/items/britishbeejourna1908lond/britishbeejourna1908lond.pdf

Medicus (February 10, February, 17 and February 24, 1910). A two-queen system: Some remarks on the adaptability for a heather district. *British Bee Journal and Bee-Keepers' Adviser* **38(1442-1444):**55-56; 64-66; 75-77. https://ia800301.us.archive.org/1/items/britishbeejourna1910lond/britishbeejourna1910lond.pdf

53   Ellis, J.M. (November 10, 1921). Notes from Gretna Green. *British Bee Journal and Bee-Keepers' Adviser* **49(2055)**:525.

54   Atkinson, M. (November 17, 1921) loc. cit.

55   Ellis, J.M. (October 13, 1921) loc. cit.

56   Ellis, J.M. (November 24, 1921). Notes from Gretna Green. *British Bee Journal and Bee-Keepers' Adviser* **49(2057)**:549.

57   Ellis, J.M. (March 9, 1922). Notes from Gretna Green. *British Bee Journal and Bee-Keepers' Adviser* **50(2072)**:105.

58   Ellis, J.M. (November 9, 1922). Notes from Gretna Green: Safe wintering. *British Bee Journal and Bee-Keepers' Adviser* **50(2107)**:547.

59   Ellis, J.M. (1923). Large hives for comb honey. *Gleanings in Bee Culture* **51(12)**:811.
      https://babel.hathitrust.org/cgi/pt?id=ucl.b4243687&view=1up&seq=625

60   Burkhardt, A.A. (December 1958). An experimental hive. *Gleanings in Bee Culture* **86(12)**:716-718, 721.
      https://archive.org/details/sim_gleanings-in-bee-culture_1958-12_86_12/page/716/mode/2up
      Vitez, P. (June 1961). My start with bees in Argentina. *Gleanings in Bee Culture* **78(6)**:428-431.
      https://archive.org/details/sim_gleanings-in-bee-culture_1961-07_89_7/page/428/mode/2up?q=TWO-QUEEN+HIVES

61   Cuello, M.A. (2005). Manejo para aumentar la producción de miel con sistema doble reina en cámara horizontal, 19pp.
      https://archive.org/details/308866809-manejo-de-colmenas-con-doble-reyna/page/12/mode/2up?q=TWO-QUEEN+hive
      https://archive.org/details/308866809-manejo-de-colmenas-con-doble-reyna_202006/page/n1/mode/2up?q=TWO-QUEEN+hive
      https://ia803206.us.archive.org/16/items/308866809-manejo-de-colmenas-con-doble-reyna_202006/308866809-Manejo-de-Colmenas-Con-Doble-Reyna.pdf

62   Gerstner, E. (February 1949). Double-hive arrangement. *Gleanings in Bee Culture* **77(2)**:80-83, 115.
      https://archive.org/details/sim_gleanings-in-bee-culture_1949-02_77_2/page/80/mode/2up?q=TWO-QUEEN+HIVES

63   Gutierrez, P.J. and Rebolledo, R.R. (2000). Comparison between a double queen and a traditional one queen system per beehive for honey production. *Agro Sur* **28(2)**:10-14.
      https://eurekamag.com/research/003/686/003686522.php

64   The recent emergence of highly insulated high density polystyrene hives confers the advantages of better thermal regulation of overwintered hives achieved by the likes of Ferris and Atkinson.

65   Nabors, R. (1 July 2015). Beekeeping topics: My two-queen hive experiment – Part I. *American Bee Journal* https://americanbeejournal.com/my-two-queen-hive-experiment/
      Nabors, R. (1 August 2015). Beekeeping topics: My two-queen hive experiment – Part II. *American Bee Journal* https://americanbeejournal.com/my-two-queen-hive-experiment-part-ii/
      Nabors, R. (January 1, 2016). Beekeeping topics: My two-queen colony hive experiment and comb honey care. *American Bee Journal* **155(9)**:1005-1006. https://americanbeejournal.com/my-two-queen-hive-experiment-and-comb-honey-care/
      Nabors, R. (January 1, 2016). Beekeeping topics: More on my new two-queen hive experiment. *American Bee Journal* **155(10)**:1125-1126.
      https://bluetoad.com/publication/?i=271638&p=77&pp=1&view=issueViewer

Nabors, R. (February 1, 2016). Beekeeping topics: My new two-queen hive design. *American Bee Journal* **156(2)**:219-220. https://ento.psu.edu/pollinators/publications/twoQueesn https://americanbeejournal.com/my-new-two-queen-hive-design/

Nabors, R. (1 March 2016). Beekeeping topics: More on my new two-queen colony. *American Bee Journal* **156(3)**:345-346. https://americanbeejournal.com/two-queen-colony/

Nabors, R. (May 1, 2016). Beekeeping topics: How to make covers for my two-queen horizontal hive. *American Bee Journal* **156(5)**:649-651. https://americanbeejournal.com/1726-2/

Nabors, R. (June 1, 2016). Beekeeping topics: Two-queen management and the different varieties of honey bees. *American Bee Journal* **156(6)**:577-578. https://americanbeejournal.com/two-queen-management-different-varieties-honey-bees/

Nabors, R. (August 1, 2016). US Department of Agriculture Products Research Report 55. Beekeeping topics: Comparing two-queen colony management methods. *American Bee Journal* **156(8)**:929-931.
http://americanbeejournal.com/comparing-two-queen-colony-management-methods/

Hesbach, W. (2015). Two queens, one hive=lots of honey. *Honey Bee Suite.* https://honeybeesuite.com/two-queens-one-hivelots-of-honey/

Hesbach,W. (2016). The horizontal two-queen system. *Bee Culture* **144(3)**:63-66. http://www.beeculture.com/the-horizontal-two-queen-system/

Wyns, D. (April 2021). Horizontal two-queen system. *Bee Culture* 44-45. https://www.beeculture.com/wp-content/uploads/2021/04/April2021_DigitalV2.pdf

66   Jones, C.L. (May 1959). A two-queen colony that pays. *Gleanings in Bee Culture* **87(5)**:276-277. https://archive.org/details/sim_gleanings-in-bee-culture_1959-05_87_5/page/276/mode/2up

67   Gouget, C.W. (July 1953). The three-queen pyramid. *Gleanings in Bee Culture* **81(7)**:395-398. https://archive.org/details/sim_gleanings-in-bee-culture_1953-07_81_7/page/394/mode/2up

68   Crane, E. (1980). Multiple queen hives and hyper hives in *Perspectives in world agriculture: Apiculture*, Chapter 10. pp.261-294. Farnham Royal, UK: Commonwealth Agricultural Bureaux. https://www.evacranetrust.org/uploads/document/b00fb1cb4f88f874217e28dcf04930fb5639e1bd.pdf

Eva Crane Gallery (1967) Ken Gray's coffin apiary in Wandoo woodland and Sid Murdoch at Manjimup operating multiqueen hives with coffin supers.
https://www.evacranetrust.org/gallery/australia

Smith, F G. (1966). *The hive.* Department of Agriculture and Food, Western Australia, Perth. Bulletin 3464, 56pp, The coffin hive p.16. https://researchlibrary.agric.wa.gov.au/cgi/viewcontent.cgi?article=1007&context=bulletins3

69   Duff, S.R. and Furgala, B. (1990). A comparison of three non-migratory systems for managing honey bees (*Apis mellifera* L.) in Minnesota: Part I Management and productivity. *American Bee Journal* **130(1)**:44-48. http://garybees.cfans.umn.edu/sites/garybees.dl.umn.edu/files/comparison_i.pdf

Furgala, B. and Sugden, M.A. (1980). Horizontal two-queen system. Unpublished working paper. Department of Entomology, University of Minnesota, 4pp.

70   Hueter, P. (23 January 1984). Multi-queen beehive. Patent application no.:490,524. https://patentimages.storage.google*Apis*.com/9a/9f/68/86e65bcc8c8556/US4546509.pdf

71   Bostok, J. and Riley, H.T. (1855). *Natural history of Pliny* Volume III, Book 11, The various kinds of insects, Chapter 4 Bees p.5. https://archive.org/details/naturalhistoryof03plin/page/4/mode/2up

72   Bostok, J. and Riley, H.T. (1855) loc. cit. Chapter 16, How honey is tested: Ericaeum,

tetralix, or sisirum, Chapter 16, p.17. https://archive.org/details/naturalhistoryof03plin/page/n37/mode/2up

73   Tarpy, D.R., Delaney, D.A. and Seeley, T. (2015). Mating frequencies of honey bee queens (*Apis mellifera* L.) in a population of feral colonies in the northeastern United States. *PLoS ONE* **10(3)**:e0118734 https://doi.org/10.1371/journal.pone.0118734

74   Culliney, T (1983). Origin and evolutionary history of the honeybees *Apis*. *Bee World* **64(1)**:29-38. DOI: 10.1080/0005772X.1983.11097905

75   Natural History Museum Los Angeles (2015). Invertebrate palaeontology. *Apis henshawi* Cockerell, 1907 (Henshaw's honey bee). https://research.nhm.org/ip/Apis-henshawi/

76   Bradbear, N. (2009). *Bees and their role in forest livelihoods:* A guide to the services provided by bees, Chapter 6. Meliponiculture of stingless bees, FAO Rome. http://www.fao.org/3/i0842e/i0842e00.htm

77   Cardinal, S. and Danforth, B.N. (2011). The antiquity and evolutionary history of social behavior in bees. *PLoS ONE* **6(6)**:e21086. https://doi.org/10.1371/journal.pone.0021086

78   Palmer, K.A. and Oldroyd, B.P. (2000). Evolution of multiple mating in the genus *Apis*. *Apidologie* **31**:235–248. https://www.Apidologie.org/articles/apido/pdf/2000/02/m0203.pdf

79   Nowak, M.A., Tarnita, C.E. and Wilson, E.O. (2010). The evolution of eusociality. *Nature* **466(7310)**:1057–1062. https://doi.org/10.1038/*Nature*09205 https://www.ncbi.nlm.nih.gov/pmc/articles/PMC3279739/

80   Hepburn, H.R. and Radloff, S.E. (1998). *Honeybees of Africa*. Springer-Verlag. Berlin, New York, 370pp, p.139.

81   Huber, F. (1806). *New observations on the natural history of bees*, Volume I by François Huber, Letters to M. Bonnet. [Letters written between 1787 and 1791] http://www.bushfarms.com/huber.htm

82   von Buttel-Reepen, H. (1900). Sind die bienen reflexmaschinen? Experiementelle beiträge zur biologie der honigbiene. *Biologisches Centralblatt* **20(1)**:1-82. Are bees reflex machines? Experimental contribution to the natural history of the honey-bee translated by Mary H. Geisler. Medina, The A.I. Root Co., 1907. 51pp, p.10. https://babel.hathitrust.org/cgi/pt?id=hvd.hn4t52;view=1up;seq=19

83   Butz, V.M. and Dietz, A. (1994). The mechanism of queen elimination in two-queen honey bee (*Apis mellifera* L.) colonies. [Nouvelles Observations Sur Les Abeilles]. *Journal of Apicultural Research* **33(2)**:87-94. https://doi.org/10.1080/00218839.1994.1110085

     Dietz, A. (1986). Theories on two-queen colonies. *American Bee Journal* **126(2)**:84, 133. Cited by Butz and Dietz (1994) loc. cit.

84   Wilson, E.O. and Hölldobler, B. (2005). Eusociality: Origin and consequences. *Proceedings of the National Academy of Sciences of the United States of America* **102(38)**:3367-3371. https://www.pnas.org/content/102/38/13367

85   Kelsall, A. (1940). A multi-queened colony of bees. *American Bee Journal* **80(4)**:170. https://archive.org/details/sim_american-bee-journal_1940-04_80_4/page/170/mode/2up

     Kosanke, M.W. (1953). Two-queen colonies. *American Bee Journal* **93(2)**:59. https://archive.org/details/sim_american-bee-journal_1953-02_93_2/page/58/mode/2up

     Farrar, C.L. (1953). Two-queen colony management. *American Bee Journal* **93(3)**:108-110, 117. https://archive.org/details/sim_american-bee-journal_1953-03_93_3/page/108/mode/2up

     Farrar, C.L. (1953). Two-queen colony management. *Bee World* **34(10)**:189-194. https://doi.org/10.1080/0005772X.1953.11094821

86   Doolittle, G.M. (1889). Scientific queen-rearing as practically applied being a method by

which the best of queen-bees are reared in perfect accord with nature's ways: for the amateur and veteran in bee-keeping 184pp, p.27. Chicago, Ills. Thomas G. Newman & Son, 923 & 925 West Madison Street. https://archive.org/details/bp_6250849

von Buttel-Reepen (1900), p.10, Spoja, J. (1953), p.195 loc. cit. and Phillips, E.F. (1915). Beekeeping: A discussion of the life of the honeybee and of the production of honey. pp.42-43. The Rural Life Sciences, Ed. Bailey, L.H. https://victoriancollections.net.au/media/collectors/51d110e42162ef12e06aa06b/items/5341f5212162ef0a845d5a79/item-media/5341f6e42162ef0a845d648d/original.pdf

87 Butler, C.G. (1954). *The World of the Honeybee*, p.60 and plate 18, p.99. Collins New Naturalist, London. Collins, St James Place, London.

Butler, C.G. (1954). The method and importance of the recognition by a colony of honeybees (*A. mellifera*) of the presence of its queen. *Transactions of the Royal Entomological Society of London* **105(pt. 2):**11-29. https://onlinelibrary.wiley.com/doi/abs/10.1111/j.1365-2311.1954.tb00773.x

Butler, C.G. (1957). The process of queen supersedure in colonies of honeybees (*Apis mellifera* L.). *Insectes Sociaux* **4(3):**211-223. https://page-one.springer.com/pdf/preview/10.1007/BF02222154

Butler, C.G. (1960). The significance of queen substance in swarming and supersedure in honey-bee (*Apis mellifera* L.) colonies. *Proceedings of the Royal Entomological Society of London.* Series A, General Entomology **35(7-9):**129-132. https://doi.org/10.1111/j.1365-3032.1960.tb00681.x Reprinted in April 2009 at *Physiological Entomology* **35(7-9):**129-132.

Neumann, P., Pirk, C.W.W., Hepburn, H.R. and Radloff, S.E. (2001). A scientific note on the natural merger of two honey bee colonies (*Apis mellifera capensis*). *Apidologie* **32(1):**113-114. https://doi.org/10.1051/apido:2001116

Simpson, J. (1958). The factors which cause colonies of *Apis mellifera* to swarm. *Insectes Sociaux* **5(1):**77-95. https://doi.org/10.1007/BF02222430

Simpson, J. (1958). The problem of swarming in beekeeping practice. *Bee World* **39(8):**193-202. https://doi.org/10.1080/0005772X.1958.11095063 https://www.tandfonline.com/doi/abs/10.1080/0005772X.1958.11095063

88 Collison, C. (2018). Venom/venom glands: A closer look. *The Australasian Beekeeper* **120(3):**46-48.

89 Doolittle, G.M. (1889) loc. cit.

90 Miller, C.C. (1911). *Fifty years among the bees*. A.I. Root Company, Medina, Ohio, Rearing queens in hive with laying queen, pp.310-312. https://archive.org/details/fiftyyearsamongb00mill

Collison, C. (2005). Do you know? Swarming and nectar. *Bee Culture* **133(4):**63-65.

Collison, C.H. (8 May and 30 June 2014). Swarm management. Publication 1817, 4pp. Extension Service of Mississippi State University, cooperating with US Department of Agriculture. https://extension.tennessee.edu/Rutherford/Documents/p1817%20swarm%20management%20by%20Dr.%20Clarence%20Collison.pdf

Collison, C.H. (2018). Swarming behavior: A closer look. *Bee Culture*. https://www.beeculture.com/a-closer-look-20/

91 Brichter, G. (August 1921). Two queens in a hive. *British Bee Journal and Bee-Keepers' Adviser* **49:**384. https://ia800203.us.archive.org/17/items/britishbeejourna1921lond/britishbeejourna1921lond.pdf

92 Hepburn, H.R. and Radloff, S.E. (1998). Chapter 5, p.140 loc. cit.

93 Claus, B. (1984). Ex Africa: Botswana, some observations on biology and behaviour. *South*

African Bee Journal **56(5):**113-116. Cited by Hepburn, H.R. and Radloff, S.E. (1998), p.140 loc. cit.

94 Mathis, M. (1952). Polygyny, temporary but constant and natural in *Apis mellifica* var. *punica* in Tunisia: Absence of mortal combat among the queens. *Comptes rendus hebdomadaires des séances de l'Académie des sciences* **234(24):**2390-2392. PMID: 12979358 [Note Mathis's reference is likely a variant of the local Tellian Honey Bee, *Apis mellifera intermissa.*] https://www.ncbi.nlm.nih.gov/pubmed/12979358

Mathis, M. (1952). Temporary (but natural and maintained) polygyny in *A. mellifera* var. *punica* in Tunisia: Queens not killing each other. *Gazette Apicole* **53(548):**230-223. Apicultural abstract 134/53. Cited by Walton, G.M. (1974) loc.cit.

95 Winston, M.L. (1980). Swarming, after swarming, and reproductive rate of unmanaged honeybee colonies (*Apis mellifera*). *Insectes Sociaux* **27(4):**391-398. https://link.springer.com/article/10.1007/BF02223731

Winston, M.L., Taylor, O.R. and Otis, G.W. (1980). Swarming, colony growth patterns, and bee management. *American Bee Journal* **120(12):**826-830.

96 Lensky, Y. and Slabezki, Y. (1981). The inhibiting effect of the queen bee (*Apis mellifera* L.) foot-print pheromone on the construction of swarming queen cups. *Journal of Insect Physiology* **27(5):**313-323. https://doi.org/10.1016/0022-1910(81)90077-9

97 Hepburn, H.R. and Radloff, S.E. (1998), p.139 loc. cit.

98 Brother Adam (1983). In search of the best strains of bees and the results of the crosses and races, 208pp. Northern Bee Books, Hebden Bridge, West Yorkshire, U.K.

99 Hogg, J.A. (2006). Colony level honey bee production: The anatomy of reproductive swarming. *American Bee Journal* **146(2):**131-135. http://www.twilightmd.com/Samples/Hogg/Hogg_Halfcomb___Publications/ABJ_2006_February.pdf

100 Killion, E.E. (1981). *Honey in the Comb.* Dadant & Sons. Inc, Carthage, Illinois.

101 Esko Suomalainan describes three types of parthenogensis: Arrhenotoky where unfertilised eggs develop pathenogenetically into males; Thelytoky where unfertilised eggs develop into females; and Deuterotokky where unfertilsed eggs develop into both sexes. Suomalainen, E. (1950). Parthenogenesis in animals. *Advances in Genetics* **3:**193-253. https://doi.org/10.1016/S0065-2660(08)60086-3

102 Goudie, F. and Oldroyd, B.P. (2014). Thelytoky in the honey bee. *Apidologie* **45(3):**306–326. https://link.springer.com/article/10.1007/s13592-013-0261-2

Oldroyd, B.P., Allsopp, M.H., Gloag, R.S., Lim, J., Lyndon A. Jordan, L.A. and Beekman, M. (2008). Thelytokous parthenogenesis in unmated queen honeybees (*Apis mellifera capensis*): Central fusion and high recombination rates. *Genetics* **180(1):**359-366. https://doi.org/10.1534/genetics.108.090415 https://www.genetics.org/content/180/1/359.full

Beekman M., Allsopp M.H., Lim J., Goudie F. and Oldroyd, B.P. (2011). Asexually produced Cape honeybee queens (*Apis mellifera capensis*) reproduce sexually. *Journal of Heredity* **102(5):**562-566. https://10.1093/jhered/esr075 https://www.researchgate.net/publication/51508237_Asexually_Produced_Cape_Honeybee_Queens_Apis_mellifera_capensis_Reproduce_Sexually

Pirk, C., Lattorff, H., Moritz, R., Sole, C., Radloff, S., Neumann, P., Hepburn, H., Crewe, R., Beekman, M., Allsopp, M., Lim, J., Goudie, F. and Oldroyd, B. (2012). Reproductive biology of the Cape Honeybee: A critique of Beekman et al. *Journal of Heredity* **103(4):**612-614. https;//doi.org/10.1093/jhered/ess007

Beekman, M., Allsopp, M.H., Lim, J., Gouldie, F. and Oldroyd, B. (2012). Response to reproductive biology of the Cape Honey Bee: A critique by Beekman et al. by Pirk et al. *Journal of Heredity* **103(4):**614-615.

doi.org/10.1093/jhered/ess008 https://academic.oup.com/jhered/article/103/4/614/1024416?login=false

103 Mackenson, O. (1943). The occurrence of parthenogenic females in some strains of honey bees. *Journal of Economic Entomology* **36(3):**465-467. https://doi.org/10.1093/jee/36.3.465

104 Butler, C.G. (1954) pp.57-58 loc. cit.

105 Smith, F.G. (1961). The races of honeybees in Africa. *Bee World* **42(10):**255-260. https://doi.org/10.1080/0005772X.1961.11096896

106 Hewitt, J. (1883). Syrian queens and fertile workers. *British Bee Journal and Bee-keepers' Record and Adviser* **11(124):**66-67. https://biodiversitylibrary.org/page/25729174

Hewitt, J. (May 1884). New races of bees: Fertile workers. *American Bee Journal* **20(19):**294-295. https://archive.org/details/sim_american-bee-journal_1884-05-07_20_19/page/294/mode/2up

A Hallamshire Beekeeper [pseudonym for John Hewitt] (1891). Qualities of the African, or Punic, bee. *American Bee Journal* **28(22):**701-702. https://www.biodiversitylibrary.org/item/77776#page/515/mode/1up Cited and refuted in detail by Baldensperger (1918) loc. cit. pp.375-376.

Hewitt, J. [A Lanarkshire Beekeeper] (July 7, 1892). Punic bees: More light wanted. *Journal of Horticulture, Cottage Gardiner and Home Farmer* **25, 3rd Series:**19. https://archive.org/details/journalofhorticu3251hogg/page/18/mode/2up

Hewitt, J. [A Lanarkshire Beekeeper] (August 4, 1892). Aparian notes: Fertile workers. *Journal of Horticulture and Cottage Gardiner* **5(3):**111-112. https://archive.org/details/journalofhorticu3251hogg/page/110/mode/2up

Hewitt, J. [A Lanarkshire Beekeeper] (August 11, 1892). Aparian notes: Faulty comb foundation. *Journal of Horticulture, Cottage Gardiner and Home Farmer* **25, 3rd Series:**133-134. https://archive.org/details/journalofhorticu3251hogg/page/132/mode/2up

Hewitt, J. [A Lanarkshire Beekeeper] (August 11, 1892). Fertile workers: Their utility.. *Journal of Horticulture, Cottage Gardiner and Home Farmer* **25, 3rd Series:**134. https://archive.org/details/journalofhorticu3251hogg/page/134/mode/2up

107 Cowan, T.W. (1891). Topics of interest: English editors and Punic bees. *American Bee Journal* **28(24):**747-749. https://www.biodiversitylibrary.org/item/77775#page/527/mode/1up

https://archive.org/details/sim_american-bee-journal_1891-12-10_28_24/page/746/mode/2up

Cowan, T. (December 17, 1891). Punic and Minorcan bees – explanation. *American Bee Journal* **28(25):**811-813. https://www.biodiversitylibrary.org/item/77775#page/575/mode/1up https://ia902609.us.archive.org/23/items/americanbeejourn28hami/americanbeejourn28hami.pdf

108 Baldensperger, P.J. (1918). Punics and parthenogenesis. *American Bee Journal* **58(11):**375-376.
https://archive.org/details/sim_american-bee-journal_1918-11_52_11/page/374/mode/2up

109 Onions, G.W. (1912). South African fertile-worker bees. *Agricultural Journal of the Union of South Africa* **3(5):**720-728.
https://journals.co.za/content/ajusa/3/5/AJA0000021_1356?fromSearch=true

Onions, G.W. (1914). South African fertile-worker bees. *Agricultural Journal of the Union of South Africa* **7:**44-46. https://journals.co.za/content/ajusa/7/1/AJA0000021_317?fromSearch=true

Jack, R.W. (1917). XXI. Parthenogenesis amongst workers of the Cape Honey Bee: Mr G.W.

Onions' experiments. *Transactions of the Royal Entomological Society of London* **64(3-4)**:396-403. https://www.biodiversitylibrary.org/item/55138#page/14/mode/1up

110 van Warmelo, D.S. (1912). South African fertile-worker bees and parthenogenesis. *Agricultural Journal of the Union of South Africa* **3**:786-789. https://journals.co.za/content/ajusa/3/6/AJA0000021_611?fromSearch=true

Nachtsheim, H. (1912). Parthenogenese, eireifung, und geschlechtsbestimmung bei der honigbiene. *Sitzungsberichte der Gesellschaft für Morphologie und Physiologie in München* **28**:22-29. Cited by Nachtsheim, H. (1913) loc. cit.

Nachtsheim, H. (1913). Cytologische studien über die geschlechtsbestimmung bei der honigbiene (*Apis mellifica* L.). *Archiv für Zellforschung* **11**:573-608. http://honeybee.drawwing.org/node/971 Cited by Oldroyd et al (2008) loc. cit.

Anderson, J. (June 1918). Laying workers which produce female offspring. *American Bee Journal* **58(6)**:192. https://ia800907.us.archive.org/21/items/americanbeejourn5859hami/americanbeejourn5859hami.pdf

https://archive.org/details/sim_american-bee-journal_1918-06_52_6/page/192/mode/2up

Gough, L.H (1928). Account of the biology of the honeybee in the Eastern Cape covering the essential features of capensis: A queenless swarm can produce a new queen. Apistischer brief aus Südafrika. *Der Bienenvater* **60**:30-32. Cited by Hepburn, S.G. and Guye, H.R. (1993). An annotated bibliography of the Cape Honeybee, *Apis mellifera capensis* Eschscholtz (Hymenoptera: Apidae). *African Entomology* **1(2)**:235-252. https://journals.co.za/content/ento/1/2/AJA10213589_64?fromSearch=true

111 Fyg, W. and Hodges, F.E.D. (1950). Can workers and queens of the honeybee be raised from unfertilised eggs? *Bee World* **31**:17-19. https://www.tandfonline.com/doi/abs/10.1080/0005772X.1950.11094622?journalCode=tbee20

Ruttner, F. and Mackensen, O. (1952). The genetics of the honeybee. *Bee World* **33(5)**:671-679. https://www.tandfonline.com/doi/abs/10.1080/0005772X.1952.11094729

Tucker, J.W. (1958). Automictic parthenogenesis in the honey bee. *Genetics* **43(3)**:299-316. www.genetics.org/content/genetics/43/3/299.full.pdf

112 von Siebold, C.T.T. (1857). *On a true parthenogenesis in moths and bees*: A contribution to the history of reproduction in animals,125 pp. Translated by William S. Dallas, F.L.S. &c. London: John van Voorst, Paternoster Row. True parthenogenesis in the honey-bee pp.38-91. https://ia802302.us.archive.org/22/items/ontrueparthenoge00sieb_0/ontrueparthenoge00sieb_0.pdf

For a broader overview of parthenogenesis see Uyenoyama, M.K. (1984). On the evolution of parthenogenesis: A genetic representation of the cost of meiosis. *Evolution* **38(1)**:87-102. https://doi.org/10.1111/j.1558-5646.1984.tb00262.x

113 Dzierżoń (1882). *Dzierzon's Rational Beekeeping or the Theory and the Practice of Dr Dzierzon of Carlsmarkt.* (Translated by H. Dieck and S. Stutterd; Edited and reviewed by Charles Nash Abbott. Houlston & Sons, Paternoster Square, Southall: Abbott Bros. from Jan Dzierzon's (1852) treatise Nachtrag zur Theorie und Praxis des neuen Bienenfreundes; oder, Einer neuen Art der Bienenzeucht mit dem günstigsten Erfolge angewendet und dargestellt von Dzierzon. Nördlingen, Beck.0 https://ia800201.us.archive.org/17/items/dzierzonsration00stutgoog/dzierzonsration00stutgoog.pdf

114 Reported by former Canberra Region Beekeepers club president Cormac Farrell in the summer of 2015-2016.

115 Hepburn, H.R. and Radloff, S.E. (1998), pp.140-141 loc. cit.

116 Hepburn, H.R. (2001). The enigmatic Cape Honey Bee. *Bee World* **82(4)**:181-191. https://www.tandfonline.com/doi/abs/10.1080/0005772X.2001.11099525

Muerrle, T.M. (January 2008), PhD Thesis. Queens, pseudoqueens and laying workers; Reproductive competition in the Cape Honeybee (*Apis mellifera capensis* Eschscholtz). https://core.ac.uk/download/pdf/11984223.pdf

117 Clauss, B. (1983). Bees and beekeeping in Botswana, Ministry of Agriculture, Gaborone. Cited in Hepburn, S.G. and Guye, H.R. (1993) loc. cit.

118 Magnum, W.A. (1 May 2009). Usurpation: A colony taken over by a foreign swarm. *American Bee Journal* **149(5)**. https://americanbeejournal.com/usurpation-a-colony-taken-over-by-a-foreign-swarm/

Magnum, W.A. (2010). The usurpation (takeover) of established colonies by summer swarms in Virginia. *American Bee Journal* **150(12):**1139-1144. Cited by Oliver, R. (2014). What's happening to the bees? – Part 2. http://scientificbeekeeping. com/whats-happening-to-thebees-part-2/

Magnum, W.A. (2012). Colony takeover (usurpation) by summer swarms: They chose poorly. *American Bee Journal* **153(1):**73-75.

Magnum, W.A. (1 January 2018). Usurpation: a colony taken over by a foreign swarm. *American Bee Journal* **158(1)**. https://americanbeekeeping.com/usurpation-colony-take-summer-swarms/

119 Schneider, S. Deeby, T., Gilley, D. and DeGrandi-Hoffman, G. (2004). Seasonal nest usurpation of European colonies by African swarms in Arizona, USA. *Insectes Sociaux* **51(4):**359-364. https://doi.org/10.1007/s00040-004-0753-1

Danka, R.G., Hellmich, R.L. and Rinderer, T.E. (1992). Nest usurpation, supersedure and colony failure contribute to Africanization of commercially managed European honey bees in Venezuela. *Journal of Apicultural Research* **31(3-4):**119-123. https://doi.org/10.1080/002188 39.1992.11101272

120 Engle, C.S. (May 1952). Automatic Demareeing. *American Bee Journal* **92(5):**197. https://archive.org/details/sim_american-bee-journal_1952-05_92_5/page/196/mode/2up

121 Demaree, G. (1892). How to prevent swarming. *American Bee Journal* **29(17):**545-546. https://archive.org/details/sim_american-bee-journal_1892-04-21_29_17/page/544/mode/2up

122 Smith, J. (1923). *Queen rearing simplified*. A.I. Root Company, Medina, Ohio, 129pp. Chapter 38 more than one queen in a hive pp.114-115. https://archive.org/details/queenrearingsimp00smit/page/114/mode/2up http://www.bushfarms.com/beesqueenrearingsimplified.htm

https://www.forgottenbooks.com/en/books/QueenRearingSimplified_10843612

123 Alexander, E.W. (April 1, 1907). Laying queens: Is it practical to have two or more in one colony during the summer season? *Gleanings in Bee Culture* **35(17):**473-474. https://babel.hathitrust.org/cgi/pt?id=umn.31951d00953180r&view=1up&seq=269

Alexander, E.W. (May 15, 1907). Queen-rearing: Some questions answered concerning the age of drones, the two-queen system and other matters. *Gleanings in Bee Culture* **35(10):**694. https://babel.hathitrust.org/cgi/pt?id=umn.31951d00953180r&view=1up&seq=398

Alexander, E.W. (April 1907). Plurality of queens in one hive, pp. 80-82 in Root, H.H. (ed) (1910). *Alexander's writings on practical bee culture*, third edition, 124pp. A.I. Root Company, Medina, Ohio. https://ia800206.us.archive.org/18/items/cu31924003065244/cu31924003065244.pdf See also: Editor of *Gleanings* (May 1, 1907). Notes on Mr E.W. Alexander and Mr E.E. Pressler, of Williamsport, Pa. on operating hives with two queens. A plurality of queens in a hive without perforated zinc or other division. *Gleanings in Bee Culture* **35(9):**617. https://www.biodiversitylibrary.org/item/74679#page/631/mode/1up

124 Alexander, E.W. (September 1,1907). A plurality of queens without perforated zinc: How the

queens are introduced: The advantages of the plural-queen system. *Gleanings in Bee Culture* **35(17)**:1136-1138.

https://babel.hathitrust.org/cgi/pt?id=umn.31951d00953180r&view=1up&seq=660

125   Alexander, E.W. (December 1, 1907). The plural queen system: All queens but one disappear at the end of the season. *Gleanings in Bee Culture* **35(23)**:1496.

https://babel.hathitrust.org/cgi/pt?id=hvd.32044106180268&view=1up&seq=566

126   Hall, C.A. (December 15, 1907). Introducing queens: A modification of the Alexander method [with a response from E.W. Alexander]. *Gleanings in Bee Culture* **35(24)**:1592. https://babel.hathitrust.org/cgi/pt?id=umn.31951d00953180r;view=1up;seq=930

127   Brother Adam (2018). *The Introduction of Queen Bees.* Hebden Bridge, West Yorkshire, U.K.

128   Alexander, E.W. (September 1,1907) loc. cit.

129   Macdonald, D.M. (October 31, 1907). American and colonial papers: Extracts and comments. *British Bee Journal and Bee-Keepers' Adviser* **35(1223)**:437-438. https://ia800201.us.archive.org/18/items/britishbeejourna1907lond/britishbeejourna1907lond.pdf

https://www.biodiversitylibrary.org/item/83074#page/449/mode/1up

Avery, G.W. (November 7, 1907). The quiet season: Some retrospective bee notes. *British Bee Journal and Bee-Keepers' Adviser* **35(1224)**:443-444. https://ia600201.us.archive.org/18/items/britishbeejourna1907lond/britishbeejourna1907lond.pdf

https://www.biodiversitylibrary.org/item/83074#page/449/mode/1up

130   The National Bee Keepers' Convention, Harrisburg (October 29-30, 1907). Plurality of queens. *Gleanings in Bee Culture* **35(22)**:1430-1432. https://babel.hathitrust.org/cgi/pt?id=umn.31951d00953180r&view=1up&seq=830

131   Editorial (May 1, 1907). A plurality of queens in a hive without perforated zinc or other division. *Gleanings in Bee Culture* **35(9)**:617.

Editorial (November 15, 1907). Plurality of queens. *Gleanings in Bee Culture* **35(22)**:1432.

Green, J.A. (May 1, 1907). Bee keeping among the Rockies: Two laying queens in a colony. *Gleanings in Bee Culture* **35(9)**:618.

132   Alexander, E.W. (May 15, 1907). Queen-rearing: Some questions answered concerning the age of drones: The two-queen system and other matters. *Gleanings in Bee Culture* **35(10)**:694.

133   Alexander, E.W. (September 15, 1908). The plural queen system: Is it profitable for bee-keepers to practice it? *Gleanings in Bee Culture* **36(18)**:1135. https://www.biodiversitylibrary.org/item/57226#page/1153/mode/1up

134   Marchant, A.B. (April 1, 1911). Raising queens above perforated zinc: Plurality of queens for one colony. *Gleanings in Bee Culture* **39(7)**:227. https://babel.hathitrust.org/cgi/pt?id=ucl.b3458208&view=1up&seq=261

135   Alexander's apiary (1921). The Alexander apiary at Delanson, New York: One of the largest apiaries in the world. *American Bee Journal* **41(9)**:385. https://ia802702.us.archive.org/19/items/americanbeejourn6061hami/americanbeejourn6061hami.pdf

Lloyd, L. (October 1, 1910). Mr Alexander's apiary of 730 colonies operated by his son Frank Alexander. *Gleanings in Bee Culture* **38(19)**:636. https://babel.hathitrust.org/cgi/pt?id=hvd.32044106180169&view=1up&seq=55&q1=frank%20alexander

136   Chambers, J.E. (September 1, 1907). Two or more laying queens in one hive: The plan not a success in the hands of the average bee-keeper. *Gleanings in Bee Culture* **35(17)**:1146-1147. https://babel.hathitrust.org/cgi/pt?id=umn.31951d00953180r;view=1up;seq=670;skin=mobile

Chambers, J.E. (December 15, 1907). Double queen colonies: More than two queens in one hive not a success, and these two must be kept separate. *Gleanings in Bee Culture* **35(24):**1582. https://www.biodiversitylibrary.org/item/74679#page/1597/mode/1up

Hand, J.E. (November 1, 1907). The plural queen system: No problem to introduce a number of queens to bees, but difficult to introduce them to each other. *Gleanings in Bee Culture* **35(21):**1385-1386. https://babel.hathitrust.org/cgi/pt?id=umn.31951d00953180r;view=1up;seq=813

Hand, J.E. (December 15, 1907). The plural queen system: No problem to introduce a number of queens to bees, but difficult to introduce them to each other. *Gleanings in Bee Culture* **35(24):**1885. https://babel.hathitrust.org/cgi/pt?id=umn.31951d00953180r;view=1up;seq=813

Hand, J.E. (January 1, 1908). The plural-queen system: Why is it more practicable with a division board-chamber than with an ordinary full-depth hive. *Gleanings in Bee Culture* **36(1):**35-36. https://babel.hathitrust.org/cgi/pt?id=uc1.b3458202;view=1up;seq=47

Hand, J.E. (February 1, 1908 continued from January 1, 1908). The two-queen system: This plan makes it possible to keep the brood chamber packed with brood during the flow: Forcing honey into the supers: Wintering two queens in one hive not desirable. *Gleanings in Bee Culture* **36(3):**155-156. https://babel.hathitrust.org/cgi/pt?id=uc1.b3458202;view=1up;seq=167

Hand, J.E. (April 15, 1908). The dual and plural queen systems: Conditions under which they may be used: Review of the whole question. *Gleanings in Bee Culture* **36(8):**507-508. https://babel.hathitrust.org/cgi/pt?id=uc1.b3458202&view=1up&seq=519

Wright, A.J. (November 1, 1907). The Alexander plan of building up weak colonies, and a modification of it: Two queens in a hive as a method of preventing swarming. *Gleanings in Bee Culture* **35(21):**1386. https://babel.hathitrust.org/cgi/pt?id=umn.31951d00953180r;view=1up;seq=814

Davenes, H. (December 15, 1907). The plural queen system: The colonies more uniformly strong at the beginning of the honey-flow: Swarming easier to control: Why only one queen is left when excluders are removed. *Gleanings in Bee Culture* **35(24):**1578-1579, 1582 https://www.biodiversitylibrary.org/item/74679#page/1594/mode/1up

Sherrod, J. (December 15, 1907). The plural-queen system: More honey than for the one-queen system. *Gleanings in Bee Culture* **35(24):**1593-1594. https://www.biodiversitylibrary.org/item/74679#page/1609/mode/1up

Bussy, E. (February 1, 1908). The plural queen system: A series of interesting experiments: Clipping the queen's stings so they can't kill eachother: Do the bees take a hand in royal combat? *Gleanings in Bee Culture* **36(3):**156-157. https://babel.hathitrust.org/cgi/pt?id=uc1.b3458202;view=1up;seq=168

Gray, J. (February 1, 1908). The plural-queen system: How an English expert looks at the question: Advantages and disadvantages. *Gleanings in Bee Culture* **36(3):**157. https://babel.hathitrust.org/cgi/pt?id=uc1.b3458202;view=1up;seq=169

Gray, J. (April 15, 1908). The plural-queen system: A review of the English and American systems: Their advantages and disadvantges as viewed from the standpoint of a travelling expert in England. *Gleanings in Bee Culture* **36(8):**505. https://babel.hathitrust.org/cgi/pt?id=uc1.b3458202&view=1up&seq=517

Joice, G.W. (April 1, 1911). Wintering a surplus of queens in one colony: The plan a success. *Gleanings in Bee Culture* **39(7):**221. https://babel.hathitrust.org/cgi/pt?id=uc1.b3458208&view=1up&seq=255

Joice, G.W. (July 15, 1911). More about wintering a surplus of queens in one colony: Is it worth a trial. *Gleanings in Bee Culture* **39(14):**436-437. https://babel.hathitrust.org/cgi/

pt?id=uc1.b3458209&view=1up&seq=100

Simmins, S. (April 15, 1908) loc.cit.

137 Robinson, T.P. (December 15, 1907). Plurality of queens in one hive: What are the advantages. *Gleanings in Bee Culture* **35(24)**:1598. https://www.biodiversitylibrary.org/item/74679#page/1614/mode/1up

Whitney, W.M. (January 1, 1908). The plural-queen system: A protest against the plan: Time and money should be spent in breeding better queens rather than striving to make a lot of poor queens live together. *Gleanings in Bee Culture* **36(1)**:36-37. https://babel.hathitrust.org/cgi/pt?id=uc1.b3458202&view=1up&seq=48

138 Wade, A. (December 2019). Doubling hives Part I – Doubled versus two-queen hives. Canberra Region Beekeepers Newsletter https://actbeekeepers.asn.au/bee-buzz-box-december-2019-doubling-hives-part-i-doubled-vs-two-queen-hives/?preview_id=6911&preview_nonce=066f240efa&preview=true&_thumbnail_id=7059

Wade, A. (February 2020). Bee Buzz Box February 2020 Doubling Hives Part II – The Wells System. – Part II The Wells System. Canberra Region Beekeepers Newsletter. https://actbeekeepers.asn.au/bee-buzz-box-february-2020-doubling-hives-part-ii-the-wells-system/?preview_id=8466&preview_nonce=f4183004ba&preview=true&_thumbnail_id=7087

139 Cruadh (December 15, 1907). Plurality of queens: Another plan for introduction: Two or more queens to a colony. *Gleanings in Bee Culture* **35(24)**:1592-1593. https://www.biodiversitylibrary.org/item/74679#page/1608/mode/1up

140 Medicus (December 17, 1908) loc. cit.

141 Ellis, J.M. (November 26, 1908) loc. cit.

Ellis, J.M. (December 31, 1908) loc. cit.

142 Strange, J.P., Garnery, L. and Sheppard, W.S. (2007). Persistence of the Landes ecotype of *Apis mellifera mellifera* in southwest France: confirmation of a locally adaptive annual brood cycle trait. *Apidologie* **38(3)**:259–267. https://doi.org/10.1051/apido:2007012 https://hal.archives-ouvertes.fr/hal-00892261/document

143 Medicus (1910) loc.cit.

144 More correctly there need be at least two entrances, one extra one being needed to allow drone flight for the additional queen.

145 Spoja, J. (1953). Observations on the operation of multiqueen colonies. *Bee World* **34(10)**:195-200. https://doi.org/10.1080/0005772X.1953.11094822

146 Kovtun, F.N. (1949). How to make and use multiple queen colonies. *Pchelovodstovo* **26(9)**:29-30. Cited by Spoja, J. (1953) and by Latif, A., Qayyum, A. and ul Haq, M. (1960) loc. cit.

Kovtun, F.N. (1950). Letter to the editorial office (multiple-queen colonies). *Pchelovodstovo* **27(2)**:112. Cited by Spoja, J. (1953) loc.cit. and by Darchen, R. and Lensky, Y. (1963) as Kovtun, F.N. (1949). Comment créer et utiliser des colonies à plusieurs reines? *Pchelovodstvo* **26**:413–41.

Kovtun, F.N. (1950). Colonics à plusicurs reines. *Pchelopodstvo* **27**:112.

147 Farrar, C.L. (1953). p.189 loc. cit.

148 Lensky, Y. and Golan, Y. (1966). Honeybee populations and honey production during drought years in a subtropical climate. *Scripta Hierosolymitana* **18**:27-42. in Studies in Agricultural Entomology and Plant Pathology, Avidov, Z. (Ed.). Jerusalem 1966 at the Magnes Press, The Hebrew University. https://catalogue.nla.gov.au/Record/2752040?lookfor=scripta%20hierosolymitana%2

149 Lensky, Y. and Darchen, R. (1963). Étude préliminaire des facteurs favorisant la création de sociétés polygynes d'*Apis mellifica*. [Preliminary study of the factors promoting the formation of polygynous societies of *Apis mellifica* var. *ligustica*.] *Annales de l'Abeille* **6(1):**69-73. https://hal.archives-ouvertes.fr/hal-00890173/document

Darchen, R. and Lensky, Y. (1963). Quelques problèmes soulevés par la création de sociétés polygynes d'abeilles. *Insectes Sociaux* **10(4):**337-357. https://link.springer.com/article/10.1007/BF02223064

Lensky, Y; Darchen, R. and Levy, R. (1970). L'agressivité des reines entreeles et des ouvrières vis-a-vis des reines lors de la création des sociétés polygynes d'*Apis mellifera* L. *Revue du Comportement Animal* **4(4):**50-52. 395/72. Cited by Pflugfelder, J., and Koeniger, N. *(2003).* Fight between virgin queens (*Apis mellifera*) is initiated by contact to the dorsal abdominal surface. *Apidologie* **34:**249-256. https://doi.org/10.1051/apido:2003016 https://hal.archives-ouvertes.fr/hal-00891776/document

Lensky, Y., Cassier, P., Rosa, S. and Grandperrin, D. (1991). Induction of balling in worker honeybees (*Apis mellifera* L.) by stress pheromone from Koschevnikov glands of queen bees: Behavioural, structural and chemical study. *Comparative Biochemistry and Physiology* – Part A: *Physiology* **100(3):**585-594. https://doi.org/10.1016/0300-9629(91)90374-L

Mel'Nik, M.I. (1951). Managing multiple queen colonies. *Pchelovodstvo* **28(9):**36-37. Referenced by Zheng, H.Q., Jin, S.H., Hu, F.L., Pirk, C.W.W. and Dietemann, V. (2009) loc. cit.

150 Chambers, J.E. (1910). Requeening by superseding the old queen in a natural way. *Gleanings in Bee Culture* **37(6):**178.

https://babel.hathitrust.org/cgi/pt?id=uc1.b3458204&view=1up&seq=374

Miller, A.C. (1913). Requeening without dequeening: Some interesting facts. *Gleanings in Bee Culture* **41(23):**851-852. https://www.biodiversitylibrary.org/item/57325#page/858/mode/1up

Miller, A.C. (1914). Why the smoke method of introducing is successful: Requeening without dequeening. *Gleanings in Bee Culture* **42(14):**537-538. https://www.biodiversitylibrary.org/item/69238#page/572/mode/1up

Hand, J.E. (1914). Requeening without dequeening. *Gleanings in Bee Culture* **42(1):**292. https://www.biodiversitylibrary.org/item/69238#page/313/mode/1up

Smedley, N. (1914). Requeening without dequeening in New Zealand: A duel between two queens. *Gleanings in Bee Culture* **42(1):**292-293. https://www.biodiversitylibrary.org/item/69238#page/313/mode/1up

151 Paleolog, J. (2001). An attempt at overwintering sting-clipped queens in multiple-queen colonies. *Journal of Apicultural Science* **45(1):**13-20. http://www.jas.org.pl:81/An-attempt-at-overwintering-sting-clipped-queens-in-multiple-queen-colonies,0,7.html

Paleolog, J., Kasperek, K. and Lipiński, Z. (2011). The psychological dimension of duels between western honeybees with blunted and non blunted sting. *Journal of Apicultural Science* **55(2):**85-95. https://www.researchgate.net/publication/286056166_the_psychological_dimension_of_duels_between_westerm_honeybee_queens_with_blunted_and_non_blunted_stings

Siuda, M., Wilde, J., Woyke, J., Jasiński, Z. and Madras-Majewska, B. (2014). Wintering reserve queens in mini-plus and 3-comb nuclei. *Journal of Apicultural Research* **50(1):**61-68. https://content.sciendo.com/downloadpdf/journals/jas/58/1/article-p61.xml

152 Zheng, H.Q., Jin, S.H., Hu, F.L., Pirk, C.W.W. and Dietemann, V. (2009). Maintenance and application of multiple queen colonies in commercial beekeeping. *Journal of Apicultural Research* **48(4):**290-295. https://www.tandfonline.com/doi/abs/10.3896/IBRA.1.48.4.10

https://www.researchgate.net/publication/229087767_Maintenance_and_application_of_multiple_queen_colonies_in_commercial_beekeeping

Zheng, H.Q., Jin, S.H., Hu, F.L. and Pirk, C.W.W. (2009). Sustainable multiple queen colonies of honey bees, *Apis mellifera ligustica*. [Colonias sostenibles de abejas *Apis mellifera ligustica* con múltiples reinas.] *Journal of Apicultural Research* **48(4)**:284-289. https://doi.org/10.3896/IBRA.1.48.4.09 https://www.tandfonline.com/doi/pdf/10.3896/IBRA.1.48.4.09?needAccess=true

Zheng, H.Q., Jin, S.H., Pirk, C.W.W., Dietemann, V., Crewe, R. and Hu, F.L. (2009). Honeybee multiple queen colonies in China. https://www.apimondia.com/congresses/2009/Technology-Quality/Symposia/Ten%20years%20of%20beekeeping%20with%20multiple-queen%20colonies%20in%20China%20-%20ZHENG%20Huo-Qing.pdf

Zheng, H.Q., Dietemann, V., Hu, F.L., Crewe, R. M. and Pirk, C.W.W. (2012). A scientific note on the lack of effect of mandible ablation on the synthesis of royal scent by honeybee queens. *Apidologie* **43(3)**:471–473. https://hal.archives-ouvertes.fr/hal-01003538/document

153 Mostafa, S.N., Magda, H.A.S., and El-Ansary, O. (2017). Effect of the multiple queens within colony on some honeybee activities, *Apis mellifera carnica* and sustainability of their colonies. *Mansoura Journal of Plant Protection and Pathology* **8(6)**:277-281. http://main.eulc.edu.eg/eulc_v5/Libraries/UploadFiles/DownLoadFile.aspx?RelatedBidID=NGQ5MGFjMzYtNWE2YS00MGIxLTk3NjctNDNjYzdmNGQ3MWVhX2l0ZW1zXzEyNDkyNTM3XzM1Mjg4OV9f&filename=8-309.pdf

Riedel, S.M. and Blum, M.S. (1972). Rapid adaptation by paired queens of the honey bee, *Apis mellifera*. *Annals of the Entomological Society of America* **65(4)**:825–829. https://doi.org/10.1093/aesa/65.4.825

154 Woyke, J. (1988). Problems with queen banks. *American Bee Journal* **128(4)**:276-278. https://www.researchgate.net/publication/256088341_1987_Problems_with_Queen_Banks

155 Szabo, T.I. (1975). Overwintering of honeybee queens 1: Maintenance of honeybee queens in solitary confinement. *Journal of Apicultural Research* **14(2)**:69-74. https://doi.org/10.1080/00218839.1975.11099805

Szabo, T.I. (1977). Overwintering of honey bee queens 2: Maintenance of caged queens in queenless colonies. *Journal of Apicultural Research* **16(1)**:41-46. https://doi.org/10.1080/00218839.1977.11099858

Wyborn, W.H. (1991). Mass storage of honey bee queens over winter. MSc Thesis, 130 pp. Simon Fraser University. https://summit.sfu.ca/system/files/iritems1/4760/b14477099.pdf

Wyborn, M.H., M.L. Winston, M.L. and P.H. Laflamme, P.H. (1993). Mass storage of honey bee (Hymenoptera: Apidae) queens during the winter. *The Canadian Entomologist* **(125)**:113-128. https://doi.org/10.4039/Ent125113-1

Gençer, H.V. (2003). Overwintering of honey bee queens *en mass* in reservoir colonies in a temperate climate and its effect on queen performance. *Journal of Apicultural Research* **42(40)**:61-64. https://www.researchgate.net/publication/247161387_Overwintering_of_honey_bee_queens_en_mass_in_reservoir_colonies_in_a_temperate_climate_and_its_effect_on_queen_performance

http://citeseerx.ist.psu.edu/viewdoc/download?doi=10.1.1.632.753&rep=rep1&type=pdf

Shehata, S.M. (1982). Long-term storage of queen honeybees in isolation. *Journal of Apicultural Research* **21(1)**:11-18. https://www.tandfonline.com/doi/abs/10.1080/00218839.1982.11100510

Reid, M. (1975). Storage of queen honeybees. *Bee World* **56(1)**:21-31. https://doi.org/10.1080/0005772X.1975.11097534

Klassen, S., Guzman, E. and Kelly, P. (2017). Banking multiple queens in colonies overwinter. *Ontario Bee Journal* January/February:12-13. Accessible as Klassen, S., Guzman, E. and Kelly, P. (2015-2016). Experimental queen overwintering study at University of Guelph (2015-2016) in Bixby, M., Guarna, M., Hoover, S.E. and Pernal, S.F. (2018). *Canadian Honey Bee Queen Bee Breeders' Reference Guide*, pp.39-40. http://honeycouncil.ca/wp-content/uploads/2018/12/FinalQueenBreederReferenceGuide2018.pdf

Harp, E.R. (1967). Storage of queen bees. *American Bee Journal* **107(7)**:250-251.

Harp, E.R. (1969). A method of holding large numbers of honey bee queens in laying conditions. *American Bee Journal* **109(9)**:340-341.

Levinsohn, M. and Lensky, Y. (1981). Long-term storage of queen honeybees in reservoir colonies. *Journal of Apicultural Research* **20(4)**:226-233. https://doi.org/10.1080/00218839.1981.11100501

156   Dietz, A., Wilbanks, T.W. and Wilbanks, W.G. (1983). Investigations on long-term queen storage in confined systems. *Apiacta* **3**:1-3. http://www.fiitea.org/foundation/files/1983/A.%20DIETZ,%20T.W.%20WILBANKS,%20W.G.%20WILBANKS.pdf

Dietz, A. (1985). Problems and prospects of maintaining a two-queen colony system in honey bees throughout the year. *American Bee Journal* **125(6)**:451-453.

157   Gary, N.E. (1963). Observations of mating behaviour in the honeybee. *Journal of Apicultural Research* **2(1)**:3-13. https://doi.org/10.1080/00218839.1963.11100050

Gary, N.E. (1966). Maintenance of isolated queen bees under laboratory conditions: A preliminary research report. *American Bee Journal* **106(11)**:412-414. Cited by Reid, M. (1975) loc. cit.

Gary, N.E., Hagedorn, H.H. and Marston, J. (1967). The behavior of mated queens when colonized in multiple queen groups without worker bees. *Apiacta* **3(4)**:9-12. http://www.fiitea.org/cgi-bin/index.cgi?sid=&zone=cms&action=search&categ_id=51&search_ordine=descriere

Gary, N.E. (1959). A study of natural and induced supersedure of queen honeybees (*Apis mellifera* L.). PhD thesis, Cornell University, Ithaca, N.Y., A.A.586/64, 145pp. https://www.researchgate.net/publication/34234849_A_study_of_natural_and_induced_supersedure_of_the_queen_honey_bee_Apis_mellifera_L

158   Szabo, T.I. and Townsend, G.F. (1974). Behavioural studies on queen introduction in the honeybee 1: Effect of the age of workers from a colony with a laying queen on their behaviour towards an introduced virgin queen. *Journal of Apicultural Research* **13(1)**:19-25. https://doi.org/10.1080/00218839.1974.11099754

Szabo, T.I. (1974). Behavioural studies on queen introduction in the honeybee 2: Effect of age and storage conditions of virgin queens on their attractiveness to workers. *Journal of Apicultural Research* **13(2)**:127-135. https://doi.org/10.1080/00218839.1974.11099768

Szabo, T.I. (1974). Behavioural studies on queen introduction in the honeybee 3: Relationship between queen attractiveness to workers and worker aggressiveness towards a queen. *Journal of Apicultural Research* **13(3)**:161-171. https://doi.org/10.1080/00218839.1974.11099774

Szabo, T.I. (1974). Behavioural studies on queen introduction in the honey bee 4: Introduction of queen models into honey bee colonies. *American Bee Journal* **114(5)**:174-176.

Szabo, T.I. and Smith, M.V. (1972). Behavioural studies on queen introduction in honeybees. 5: Behavioural relationship between pairs of queens without worker attendance. *Proceedings of the Royal Society of Ontario* **103**:87-96. Findings cited by Pflugfelder, J. and Koeniger, N. (2003). Fight between virgin queens (*Apis mellifera*) is initiated by contact to the dorsal abdominal surface. *Apidologie* **34(3)**:249-256. https://hal.archives-ouvertes.fr/hal-00891776/document

Szabo, T.I. (1977). Behavioural studies of queen introduction in the honeybee 6: Multiple queen introduction. *Journal of Apicultural Research* **16(2):**65-83. https://doi.org/10.1080/00 218839.1977.11099865

Waikakul, Y. (April 1979). Behavioral studies of virgin queen honey bees. MSc Thesis, The University of Manitoba, Winnipeg, Manitoba. https://www.google.com/ search?q=Waikakul%2C+Y.+(May+1973).+Behavioral+studies+of+virgin+queen+honey+ bees&oq=Waikakul%2C+Y.+(May+1973).+Behavioral+studies+of +virgin+queen+honey+bees&aqs=chrome..69i57.2013j0j7&sourceid=chrome&ie=UTF-8

159 Farrar, C.L. (1953) loc. cit.

160 Cale, G.H. (1922). The relation of queens to seasonal management. *American Bee Journal* **62(2):**61-63. https://archive.org/details/sim_american-bee-journal_1922-02_56_2/page/60/ mode/2up

161 Buntich, A. (2021). Erie Beekeeping webpage. https://erie*Beekeeping*.com/content/clayton-leon-farrar

162 Crane, E. (2003). *Making a bee-line*, 324pp. International Bee Research Association in Miksha, R. (August 2, 2018). How to predict the honey flow. *Bad Beekeeping Blog.* https:// badbeekeepingblog.com/2018/08/02/how-to-predict-the-honey-flow/

163 Wade, A. (2019). Building bees for the honey flow. Presentation to Canberra Regional Beekeepers, 18 September 2019. https://drive.google.com/file/d/1ZNLo8xAV5b6kRD4klC D5lIUhBTLPxHaq/view

164 Farrar, C.L. (August 1958) loc. cit.

165 Farrar, C.L. (January 1, 1931). A measure of some factors affecting the development of the honeybee colony, PhD Thesis, Massachusetts State College, 139pp. https://scholarworks. umass.edu/dissertations_1/889

Farrar, C.L. (1932). The influence of the colony's strength on brood-rearing. *Annual Report of the Beekeepers' Association Ontario*, 1930 and 1931. Ontario Department of Agriculture: pp.126-130.

Farrar, C.L. (1937). The influence of colony populations on honey production. *Journal of Apicultural Research* **54(12):**945-954. https://naldc.nal.usda.gov/download/IND43969007/ PDF

Farrar, C.L. (1944). Productive management of honeybee colonies in the Northern States. Circular No. 702, 2 July 1944, Washington, D.C. United States Department of Agriculture 28pp. https://archive.org/details/productivemanage702farr http://www.nnjbees.org/wp-content/uploads/2016/10/Productive-management-of-honeybee-colonies-Farrar.pdf https://babel.hathitrust.org/cgi/pt? id=uiug.30112019274536

Farrar, C.L. (1946). Productive management of honey-bee colonies, Entomology Research Division No. E-693, revision to USDA Agricultural Research Service *Department of Agriculture Bulletin* No. 702, 1946 in cooperation with the Wisconsin Agricultural Experiment Station. Published online by USDA at *Beesource* (2019). https://beesource.com/ resources/usda/productive-management-of-honey-bee-colonies/ https://beesource.com/ resources/usda/productive-management-of-honey-bee-colonies/productive-management-of-honey-bee-colonies-support-material/

Farrar, C.L. (1953) loc. cit.

Farrar, C.L. (April 1954). The Farrar plan. *American Bee Journal* **94(4):**129-130. https:// archive.org/details/sim_american-bee-journal_1954-04_94_4/page/128/mode/2up

Farrar, C.L. (1968) loc. cit.

Tegart, D. (1984). Two-queen hive management using package bees in the Peace River area,

Alberta, Canada. *Bee World* 65(2):80-84. http://www.nmhoney.com/nmhoney/Sub_Files/Two%20Queen%20Hive%20Management%20BW652pp80-84.pdf

Donaldson-Matasci, M.C., DeGrandi-Hoffman, G. and Dornhaus, A. (2013). Bigger is better: Honeybee colonies as distributed information-gathering systems. *Animal Behaviour* **85(3):**585-592. https://www.ncbi.nlm.nih.gov/pmc/articles/PMC4511854/

166 Gerdts, J., Dewar, R.L., Finstrom, M.S., Edwards, T. and Angove, M. (2018). Hygienic behaviour selection via freeze-killed honey bee brood not associated with chalkbrood resistance in eastern Australia. *PLoS ONE* **13(11):**e0203969. https://doi.org/10.1371/journal.pone.0203969

Gerdts, J. (2014). Hygienic behaviour in 50 AI breeder colonies of the Australian queen bee breeding program, Queensland Australia, 12pp. https://honeybee.org.au/wp-content/uploads/2014/12/AQBBP-HygienicTestingReport_BeeSci2014FINAL.pdf

167 Cale, G.H. (June 1952). The effect of the two-queen system on the harvest: An interview with Dr C.L. Farrar. *American Bee Journal* **92(6):**236-237. https://archive.org/details/sim_american-bee-journal_1952-06_92_6/page/236/mode/2up

168 Dugat, M. (1946). La ruche gratte ciel á plusieurs reines: Une nouvelle methode d'apiculture intensive. 3rd edition published by Marlieux, Abbaye de N.-D. des Dombes, 1947. https://catalog.hathitrust.org/Record/009058161

Dugat, M (1948). *Skyscraper hive* [Ruche gratte-ciel]. *Au Bon Meil* [with good honey] (2020) extract from French Hunter No. 600, p.132. https://www.aubonmiel.com/ruche-gratte-ciel/

169 Crane, E. (1980). Multiple queen hives and hyper hives in *Perspectives in world agriculture: Apiculture*, Chapter 10. pp.261-294. Farnham Royal, UK: Commonwealth Agricultural Bureaux. https://www.evacranetrust.org/uploads/document/b00fb1cb4f88f874217e28dcf04930fb5639e1bd.pdf

170 Dugat, M. (1948). *The Skyscraper Hive.* Authorized translation of La Ruche Gratteciel à Plusiers Reines by Norman C. Reeves. Faber and Faber, 24 Russell Square, London. https://www.abebooks.com/servlet/SearchResults?an=dugat+father&bi=h&ds=5&n=100121503&sortby=1&tn=skyscraper+hive&cm_sp=mbc-_-ats-_-filter

171 Butler, C.G. (17 July 1948). The skyscraper hive. *Nature* **162(4107):**87. https://doi.org/10.1038/162087c0

172 Farrar, C.L. (October 1936). Two-queen vs. single-queen colony management. *Gleanings in Bee Culture* **64(10):**593-596. https://archive.org/details/sim_gleanings-in-bee-culture_1936-10_64_10/page/592/mode/2up

Farrar, C.L. (May 1946). Two-queen colony management. United States Department of Agriculture. Agricultural Research Administration, Bureau of Entomology and Plant Quarantine, 14pp. http://ufdc.ufl.edu/AA00026061/00001/14j https://archive.org/stream/twoquyma00unit/twoquyma00unit_djvu.txt https://www.scribd.com/document/313855449/Two-queen-colony-managemet http://www.immenfreunde.de/docs/2queen.pdf https://ufdc.ufl.edu/AA00026061/00001

Farrar, C.L. (1953) loc. cit.

Farrar, C.L. (August 1958). Two-queen-colony management for honey production. ARS-33-48 USDA Agricultural Research Service. https://archive.org/details/twoqueencolonyma48farr/page/n1/mode/2up

Schaefer, C.W. and Farrar, C.L. (1941). The use of pollen traps and pollen supplements in developing honey bee colonies. US Bureau of Entomology and Plant Quarantine No. E-531. Cited by Haydak, M.H. and Tanquary, M. (1943) p.7 loc. cit.

Schaefer, C.W. and Farrar, C.L. (1946). The use of pollen traps and pollen supplements in developing honeybee colonies. US Bureau of Entomology and Plant Quarantine, E531, rev. 13pp. Cited by Moeller, F.E. (1980 USDA) loc. cit.

Chaudry, M. and Johansen, C.A. (1971). Management practices affecting efficiency of the honey bee *Apis mellifera* (Hymenoptera: Apidae). Scientific paper no. 3674, Washington Agricultural Experiment Station, Washington State University. *Melanderia* **6:**32pp. downloaded at CSIRO Black Mountain Library (AS:PE) P 595.7(73) Me v.6 (1971).

Haydak, M.H. and Tanquary, M. (June 1943). Pollen and pollen substitutes in the nutrition of the honeybee. *Technical Bulletin* **160**, 23pp. Division of Entomology and Economic Zoology, University of Minnesota, Agricultural Experiment Station. https://conservancy.umn. edu/bitstream/handle/11299/204094/mn1000_agexpstn_tb_160.pdf?sequence=1

Kornely, R. (1969). Care of the two-queen-colony in spring. *Gleanings in Bee Culture* **97(1):**16-20.

173 Moeller, F.E. (1961). The relationship between colony populations and honey production as affected by honey bee stock lines. US Department of Agriculture, Agricultural Research Service, Production Research Report 55, 24pp. https://babel.hathitrust.org/cgi/pt?id=uc1. aa0001977776;view=1up;seq=2

174 Crane, E. (1954). An American bee journey. *Bee World* **35(7):**125-137. https://doi.org/ 10.1080/0005772X.1969.11097261 https://www.evacranetrust.org/uploads/document/ ab43420d3fa91f1e2ee69e8e63dc32907c536a85.pdf

175 Cale, G.H. (June 1952) loc. cit.

176 Cale, G.H. (May 1952). All around the bee yard. *American Bee Journal* **92(5):**208-209. https://archive.org/details/sim_american-bee-journal_1952-05_92_5/page/208/mode/2up

177 The horizontal Langstroth hive differs from the popular free comb-forming Kenyan and Tanzanian top bar hives in accomodating standard full depth Langstroth frames.

178 Adair, D.L. (May 1872). The new idea bee hive. *American Bee Journal* **7(11):**253. https:// archive.org/details/sim_american-bee-journal_1872-05_7_11/page/252/mode/2up

https://www.biodiversitylibrary.org/item/65798#page/263/mode/1up

Gallup, E. (March 1872). A little plain talk. *American Bee Journal* **7(3):**207-208. https:// www.biodiversitylibrary.org/item/65798#page/217/mode/1up

Adair, D.L. (May 1873). New idea hive, with frames. *American Bee Journal* **8(11):**250. https://archive.org/details/sim_american-bee-journal_1873-05_8_11/page/250/mode/2up

Adair, D.L. (June 1874. Adair talks about Novice. *American Bee Journal* **8(6):**123-125. https://www.biodiversitylibrary.org/item/66213#page/89/mode/1up

Gallup, E. (June 1874). Gallup's new idea hive and its advantages. *American Bee Journal* **8(6):**133-134. https://www.biodiversitylibrary.org/item/66213#page/133/mode/1up

Doolittle, G.M. (September 1, 1897). Answers to seasonable questions: Long hives vs. tiering up. *Gleanings in Bee Culture* **25(17):**634-635. https://www.biodiversitylibrary.org/ item/69195#page/571/mode/1up

179 Gallup, E. (July 1872). Gallup on one-story hives. *American Bee Journal* **8(1):**12. https:// archive.org/details/sim_american-bee-journal_1872-07_8_1/page/12/mode/2up

Gallup, E. (March 1873). Gallup's explanations to Wurster. *American Bee Journal* **8(9):**215. https://archive.org/details/sim_american-bee-journal_1873-03_8_9/page/214/mode/2up

Argo, R.M. (May 1873). Non-swarmers. *American Bee Journal* **8(11):**249-250. https:// archive.org/details/sim_american-bee-journal_1873-05_8_11/page/248/mode/2up

Argo, R.M. (March 1873). Review of December Journal. *American Bee Journal* **8(9):**211-212. https://archive.org/details/sim_american-bee-journal_1873-03_8_9/page/210/ mode/2up

180   Eccentric (January 1875). Eccentric. *American Bee Journal* **11(1)**:11-12. https://archive.org/details/sim_american-bee-journal_1875-01_11_1/page/10/mode/2up

Novice (August 1874). Novice's answer. *American Bee Journal* **8(8)**:177. https://www.biodiversitylibrary.org/item/66213#page/175/mode/1up

Argo, R.M. (January 1875). Remarks on Eccentric. *American Bee Journal* **11(1)**:45-46. https://www.biodiversitylibrary.org/item/66213#page/333/mode/1up

Taylor, B.L. (May 1875). How I wintered. *American Bee Journal* **11(5)**:115-116. https://www.biodiversitylibrary.org/item/66213#page/400/mode/1up

181   Poppleton, O.O. (January 1, 1898). The long-idea hive: History of it, and some corrections. *Gleanings in Bee Culture* **26(1)**:13-14. https://archive.org/details/sim_gleanings-in-bee-culture_1898-01-01_26_1/page/12/mode/2up

Poppleton, O.O. (January 1899). Question box: Best size of hive for beginners. *American Bee Journal* **39(14)**:12.

Eatinger, F. (April 13, 1899). The long hive and its history. *American Bee Journal* **39(15)**: 227-228. https://www.biodiversitylibrary.org/item/72953#page/231/mode/up

Root, E.R. (January 1900). The long-ideal hive. *American Bee Journal* **41(3)**: 42-43. https://www.biodiversitylibrary.org/item/61429#page/7/mode/1up

Eds – Bibliographical (January 1901). Mr 0.0. Poppleton of Florida. *American Bee Journal* **41(4)**:57-58. https://archive.org/details/sim_american-bee-journal_1901-01-24_41_4/page/56/mode/2up

Eds (January 1901). O.O. Poppleton. *Gleanings in Bee Culture* **24(2)**:54-55. https://archive.org/details/sim_gleanings-in-bee-culture_1901-01-15_29_2/page/54/mode/2up

Eds – Weekly Budget (January 1901). *American Bee Journal* **41(5)**:68. https://archive.org/details/sim_american-bee-journal_1901-01-31_41_5/page/68/mode/2up

Ed (February1904). Long-ideal hives—Doolittle's experience. *American Bee Journal* **44(7)**:115. https://archive.org/details/sim_american-bee-journal_1904-02-18_44_7/page/114/mode/2up

Mitchell, S.H. (November 1906). The long-idea hive again: How to prevent swarming and secure the largest amount of white comb honey. *Gleanings in Bee Culture* **34(22)**:1435-1436. https://archive.org/details/sim_gleanings-in-bee-culture_1906-11-15_34_22/page/1434/mode/2up

Muth, F.W. (October 1907). Bee keeping in Florida: A visit at the home of O.O. Poppleton. *Gleanings in Bee Culture* **35(19)**:1263-1264. https://archive.org/details/sim_gleanings-in-bee-culture_1907-10-01_35_19/page/1262/mode/2up

Root, E.R. (September 1908). Migratory beekeeping: How O.O. Poppleton practices the scheme by means of gasoline-launches on the Indian River, Fla.: The long-idea hive. *Gleanings in Bee Culture* **36(18)**:1118. https://archive.org/details/sim_gleanings-in-bee-culture_1908-09-15_36_18/page/1118/mode/2up

Baldwin, E.G. (June 1911). Bee-keeping in Florida: Some representative bee-men of Florida. *Gleanings in Bee Culture* **39(12)**:363-365. https://archive.org/details/sim_gleanings-in-bee-culture_1911-06-15_39_12/page/362/mode/2up

Baker, E. (October 1914). Long idea hives. *American Bee Journal* **54(10)**:336-337. https://www.biodiversitylibrary.org/item/57328#page/342/mode/1up

Root, E.R. (February 1915). The editor's pleasure-trip down the Indian River, Florida: An interview with O.O. Poppleton, the famous bee-king of southeast Florida: The long idea hive. *Gleanings in Bee Culture* **43(4)**:143-148. https://archive.org/details/sim_gleanings-in-bee-culture_1915-02-15_43_4/page/142/mode/2up

Poppleton, O.O. (July 1915). The long idea hive; some further details supplementing the editor's article on the same subject, Feb. 15. *Gleanings in Bee Culture* **43(14):**575-577. https://www.biodiversitylibrary.org/item/57118#page/6/mode/1up

182   Muth-Rasmussen, W.M. (April 1900). Is it long idea or long ideal hive? *American Bee Journal* **41(14):**212-213. https://archive.org/details/sim_american-bee-journal_1900-04-05_40_14/page/212/mode/2up *Addendum* (May 1, 1900) p.286 Long ideal hive satisfactory.

183   Root, E.R. (May 1920). Long idea hive again: Its value to the queen breeder in control of swarming one of its excellent features. *Gleanings in Bee Culture* **48(6):**268-271.
https://archive.org/details/sim_gleanings-in-bee-culture_1920-05_48_5/page/268/mode/2up

Root, E.R. (June 1920). Extensive queen-breeding: Some big and little tricks of the trade useful to honey-producers as well as to queen-breeders. *Gleanings in Bee Culture* **48(6):**332-334. https://archive.org/details/sim_gleanings-in-bee-culture_1920-06_48_6/page/332/mode/2up

Root, E.R. (January 1918). Inexpensive wintering: An ordinary hire packed inside of a long idea hive: Also a regular hive on end: Inside of three supers. *Gleanings in Bee Culture* **46(1):**14-17. https://archive.org/details/sim_gleanings-in-bee-culture_1918-01_46_1/page/14/mode/2up

Root, E.R. (September 1917). Gleaned by asking. *Gleanings in Bee Culture* **46(9):**708. https://archive.org/details/sim_gleanings-in-bee-culture_1917-09_45_9/page/708/mode/2up

184   Reeder, F.E. (January 1918). 250 long idea hives: Some of the advantages of hives capable of horizontal expansion compared to standard hives. *Gleanings in Bee Culture* **46(1):**16-17. https://archive.org/details/sim_gleanings-in-bee-culture_1918-01_46_1/page/16/mode/2up

185   Eckert, J.E. (1937). The duo or two-queen hive. *Gleanings in Bee Culture* **65(3):**137. https://archive.org/details/sim_gleanings-in-bee-culture_1937-03_65_3/page/136/mode/2up

Dayell, M.J. (September 1948). A talk to beekeepers: No royal road to success. *Gleanings in Bee Culture* **76(9):**571-573. https://archive.org/details/sim_gleanings-in-bee-culture_1948-09_76_9/page/570/mode/2up

Albrecht, A (1950). Adventure and experiment in the apiary. *Gleanings in Bee Culture* **78(6):**337-338, 383. https://archive.org/details/sim_gleanings-in-bee-culture_1950-06_78_6/page/344/mode/2up

https://archive.org/details/sim_gleanings-in-bee-culture_1950-06_78_6/mode/2up

186   Bell, G.R. (1952). W. Leroy Bell. *American Bee Journal* **92(5):**186. https://archive.org/details/sim_american-bee-journal_1952-05_92_5/page/186/mode/2up

187   Farrar, C.L., Miller, L.F., Dunham, W.E., Schaefer, E.A., Holzberlein, J. Jr., and Cale, G.H. (April 1954) loc. cit.

188   Gilbert, C.H. (1938). The two-queen hive. *Gleanings in Bee Culture* **66(7):**417-419. Medina, Ohio. https://archive.org/details/sim_gleanings-in-bee-culture_1938-07_66_7/page/416/mode/2up

Gilbert, C.H. (July 1940). The two-queen hive and commercial honey production. *University of Wyoming Agricultural Research Station Bulletin* **239:**3-15, Wyoming. https://mountainscholar.org/bitstream/handle/20.500.11919/335/AESB_1940_Bulletin_239.pdf?sequence=1&isAllowed=y https://repository.uwyo.edu/cgi/viewcontent.cgi?referer=https://www.google.com.au/&httpsredir=1&article=1187&context=ag_exp_sta_bulletins

Lanyi, S. (1939). Doubled honey-earning by a new two-queen system. *Bee World* **20(11):**123-125. https://doi.org/10.1080/0005772X.1939.11093925

Harvey, M. (2009). Kerkhof hives: Advantages of a two-queen set up. *Bee Culture* **37(4):**65-66. https://discovery.csiro.au/primo-explore/fulldisplay?docid=TN_proquest214713243&context=PC&vid=CSIRO&lang=en_US&search_scope=All&adaptor=primo_central_multiple_fe&tab=default&query=any,contains,Harvey,%20M.%20(2009).%20%20Kerkhof%20hives:%20Advantages%20of%20a%20two-queen%20set%20up.&offset=0

Goolsbey, A. (2001). 2-Queen system. *Bee Culture* **129(5):**24-29. https://discovery.csiro.au/primo-explore/fulldisplay?docid=TN_proquest214712867&context=PC&vid=CSIRO&lang=en_US&search_scope=All&adaptor=primo_central_multiple_fe&tab=default&query=any,contains,Goolsbey,%20A.%20(2001).%20%202-Queen%20system.&offset=0

Labesque, S. (2004). 2 Queener. *Bee Culture* **132(4):**31-34. https://discovery.csiro.au/primo-explore/fulldisplay?docid=TN_proquest214704962&context=PC&vid=CSIRO&lang=en_US&search_scope=All&adaptor=primo_central_multiple_fe&tab=default&query=any,contains,Labesque,%20S.%20(2004).%20%202%20Queener&offset=0

189 Roberts, W.C. and Mackensen, O. (November 1951). Breeding improved honey bees V: Production of hybrid queens. *Gleanings in Bee Culture* **79(11):**664-667. https://archive.org/details/sim_gleanings-in-bee-culture_1951-11_79_11/page/664/mode/2up?q=TWO-QUEEN+hive
https://archive.org/details/sim_gleanings-in-bee-culture_1951-11_79_11/page/664/mode/2up

190 Victor Croker and David Leehumis (pers. comm.) from Australian Honeybee employ this technique for introducing cells in the first stage of a sophisticated program to requeen their apiaries.

191 Szabo, T.I. (1982). Requeening honeybee colonies with queen cells. *Journal of Apicultural Research* **21(40):**208-211. https://doi.org/10.1080/00218839.1982.11100543

192 Dunham, W.E. (January 1939). Modified two-queen system: Especially adapted to regions where white Dutch, Alsike, sweet clover and alfalfa are prevalent. *Gleanings in Bee Culture* **67(1):**211-212. https://archive.org/details/sim_gleanings-in-bee-culture_1939-01_67_1/page/10/mode/2up

Wurz, A.S. (May 1942). Two-queen system. *Gleanings in Bee Culture* **72(5):**281. https://archive.org/details/sim_gleanings-in-bee-culture_1942-05_70_5/page/280/mode/2up

Dunham, W.E. (May 1943). Versatility of the modified two-queen system. *Gleanings in Bee Culture*. **71(5):**264-268. https://archive.org/details/sim_gleanings-in-bee-culture_1943-05_71_5/page/264/mode/2up

Dunham, W.E. (1943). The modified two-queen system. *American Bee Journal* **83(5):**192-194. https://archive.org/details/sim_american-bee-journal_1943-05_83_5/page/192/mode/2up

Dunham, W.E. (1944). Bees: Maintenance of colonies, control of colony population for honey production and pollination. *Ohio State University Extension Bulletin* **254:**2-32. Cited by Eckert, J.E. and Shaw, F.R. (1960) loc. cit.

Dunham, W.E. (March 1947). Modified two-queen system for honey production. *Bulletin of the Agricultural Extension Service*, the Ohio State University **281:**1-16. https://scholar.google.com.au/scholar?hl=en&as_sdt=0%2C5&q=Dunham%2C+W.E.+%281948%29.++Modified+two-queen+system+for+honey+production.&btnG=

Dunham, W.E. (May 1948). Modified two-queen system for honey production. [Part of

Bulletin No. 281, issued March 1947, by Agricultural Extension Service, The Ohio State University, Columbus, Ohio.] *Gleanings in Bee Culture* **76(5):**277-281. https://archive.org/details/sim_gleanings-in-bee-culture_1948-05_76_5/page/276/mode/2up

Dunham, W.E. (April 1951). The Ohio modified two-queen system. *Gleanings in Bee Culture* **79(4):**212-214. https://archive.org/details/sim_gleanings-in-bee-culture_1951-04_79_4/page/212/mode/2up

Dunham, W.E. (1953). The modified two-queen system for honey production. *American Bee Journal* **93(3):**111-112. https://archive.org/details/sim_american-bee-journal_1953-03_93_3/page/110/mode/2up

Dunham, W.E. (April 1954). Dunham's modified two-queen plan. *American Bee Journal* **94(4):**130-.131. https://archive.org/details/sim_american-bee-journal_1954-04_94_4/page/130/mode/2up

193 Eckert, J.E. and Shaw, F.R. (1960). *Beekeeping: Successor to Beekeeping* by Everett F. Phillips, MacMillan Publishing Co. Inc., New York, 546 pp.

Eckert, J.E. and Shaw, F.R. (1969). *Beekeeping.* The Macmillan Company, New York. https://archive.org/stream/in.ernet.dli.2015.140016/2015.140016.Beekeeping-Successor-To-Beekeeping_djvu.txt

194 Eckert, J.E. and Shaw, F.R. (1969). p.217 loc. cit.

195 Wedmore, E.B. (1945). *A manual of beekeeping for English-speaking beekeepers.* 2nd Revised edition, republished 1979 by Bee Books New & Old, Steventon, Hampshire as *A manual of beekeeping,* 2nd Edn revised, pp.294-295, 312. Butler and Tanner Ltd, Frome and London.

196 Holzberlein, J.W. (May 1952). Swarm prevention—not swarm control. *American Bee Journal* **92(5):**195–196. https://archive.org/details/sim_american-bee-journal_1952-05_92_5/page/194/mode/2up

Newman, I.L. (May 1952). Swarm control with nucleus system. *American Bee Journal* **92(5):**194-195. https://archive.org/details/sim_american-bee-journal_1952-05_92_5/page/194/mode/2up

Holzberlein, J.W. (March 1953). Getting started with two-queen management. *American Bee Journal* **93(3):**114-115. https://archive.org/details/sim_american-bee-journal_1953-03_93_3/page/114/mode/2up

Schaefer, E.A. (April 1954). The Schaefer two-queen swarm control method. *American Bee Journal* **94(4):**131-132. https://archive.org/details/sim_american-bee-journal_1954-04_94_4/page/130/mode/2up

Holzberlein, J.W. (April 1954). Another way to start two-queen management. *American Bee Journal* **94(4):**131-132. https://archive.org/details/sim_american-bee-journal_1954-04_94_4/page/130/mode/2up

Holzberlein, J.W. (1955). Some whys and hows of two-queen management. *Gleanings in Bee Culture* **83(6):**344-347. https://archive.org/details/sim_gleanings-in-bee-culture_1955-06_83_6/page/344/mode/2up

197 Miller, L.F. (1953). Crop insurance with two queens. *American Bee Journal* **93(3):**113, 117. https://archive.org/details/sim_american-bee-journal_1953-03_93_3/page/112/mode/2up

Miller, L.F. (April 1954). Two queens to retrieve weak colonies: *American Bee Journal* **94(4):**130. https://archive.org/details/sim_american-bee-journal_1954-04_94_4/page/130/mode/2up

Cale, G.H. (April 1954). Reversal, separation, and reunion. *American Bee Journal* **94(4):**132. https://archive.org/details/sim_american-bee-journal_1954-04_94_4/page/132/mode/2up

198 Latif, A. and Hussain, M. (1955). Double-queen colonies of *Apis indica* F. at Sialkot.

Proceedings of the 7th Pakistan Scientific Conference, Pakistan.

Latif, A., Qayyum, A. and ul Haq, M. (1956). Two-queen system in bee-hives of *Apis indica* F. at Lyallpur in *Proceedings of the 7th Pakistan Scientific Conference, Pakistan.*

Latif, A., Qayyum, A. and ul Haq, M. (1956). Two-queen system in relation to colony development. In *Proceedings of the 7th Pakistan Scientific Conference, Pakistan.*

Latif, A., Qayyum, A. and ul Haq, M. (1960). Multiple and two-queen colonies of *Apis indica* F. *Bee World* **41(8):**201-209. https://doi.org/10.1080/0005772X.1960.11096796

Singh, J. (1953). Two queens in a colony. *Indian Bee Journal* **3(5/6):**56-57. Referenced through Latif, A., Qayyum, A. and ul Haq, M. (1960) loc. cit.

Zhang, Q.K. (1994). Discussion on the superior characteristics of two-queen colonies in *Apis cerana cerana. Apicultural Science and Technology* **5:**2. Cited by Hepburn, H.R. and Radloff, S.E. (eds) (2011). *Honeybees of Asia.* p.655. Hepburn, H.R. Springer-Verlag Berlin Heidelberg.

199 Wafa, A.K. (1956). Two-queen colonies for a plentiful yield of honey, safe wintering, means of propagation and of swarming control. *Faculty of Agriculture Cairo University Bulletin* 98. 22pp. (Apicultural Abstracts 196/58).

200 Haydak, M.H. and Dietz, A. (1967). Two-queen colonies, requeening and increase. *American Bee Journal* **107(5):**171-172. Cited by Forster, I.W. (1972) loc. cit. and by Butz, V.M. and Dietz, A. (1994). The mechanism of queen elimination in two-queen honey bee (*Apis mellifera* L.) colonies. [Nouvelles Observations Sur Les Abeilles]. *Journal of Apicultural Research* **33(2):**87-94. https://doi.org/10.1080/00218839.1994.11100855 and by Wallrebenstein (1958). My contribution to the multiple-queen system (Mein beitrag, um mermutterverfahren). XVII International Beekeeping Congress, Bologna-Roma, pp.277-286.

201 Farrar, C.L., Dunham, W.E., Miller, L.F. and Holzberlein, J.W. (March 1953). Spotlight: Two-queen management. *American Bee Journal* **93(3):**107107-115). https://archive.org/details/sim_american-bee-journal_1953-03_93_3/page/107/mode/1up

202 Farrar, C.L., Miller, L.F., Dunham, W.E., Schaefer, E.A., Holzberlein, J. Jr., and Cale, G.H. (April 1954). Panel for April: How to use two queens for automatic requeening, swarm control and crop increase. *American Bee Journal* **94(4):**128-132. https://archive.org/details/sim_american-bee-journal_1954-04_94_4/page/128/mode/2up

203 Holzberlein, J.W. (May 1952) loc. cit.

204 Holzberlein, J.W. (March 1953 and April 1954) loc. cit.

205 Holzberlein, J.W. (March 1953) loc. cit.

206 Holzberlein, J.W. (1955) loc.cit.

207 Schaefer, H.A. (1943). Two-queen system. *American Bee Journal* **83(6):**234-235. https://archive.org/details/sim_american-bee-journal_1943-06_83_6/page/234/mode/2up

208 Banker, R. (1968). A two-queen method used in commercial operations. *American Bee Journal* **108(5):**180-182. Republished as Banker R. (1968). A two-queen method used in commercial operations. *Apiacta* **2:**1-4. http://www.fiitea.org/cgi-bin/index.cgi?sid=&zone=cms&action=search&categ_id=53&search_ordine=descriere

Banker, R. (1979). Part B. Two-queen colony management in *The Hive and the Honey Bee,* Dadant & Sons, Hamilton, Illinois, Chapter XII, pp.404-410, 412.

Cale, G.H., Banker, R. and Powers, J. (1975). Management for honey production in *The Hive and the Honey Bee,* Dadant & Sons, Hamilton, Illinois, Chapter XII, pp.355-403, 412.

209 Davis, J.L. (1908). Queen killed by a rival queen. *Gleanings in Bee Culture* **36(20):**1259-1260. https://babel.hathitrust.org/cgi/pt?id=uma.ark:/13960/t84j0sg9w;view=1up;seq=1277

Skirkyavichyus, A. (1965) Can two queens live together? *Pchelovodstvo* **85(6):**16-18. Cited

by Reid, M. (1975). Storage of queen honeybees. *Bee World* **56(1):**21-31. https://doi.or
g/10.1080/0005772X.1975.11097534, by Forster, I.W. (1972) loc. cit. and by Gary, N.E.,
Hagedorn, H.H. and Marston, J. (1967). The behavior of mated queens when colonized in
multiple queen groups without worker bees. *Apiacta* **3(4):**9-12. http://www.fiitea.org/cgi-
bin/index.cgi?sid=&zone=cms&action=search&categ_id=51&search_ordine=descriere

210   Crane, E. (1966). Canadian bee journey. *Bee World* **47:(4):**132-148. https://doi.org/1
0.1080/0005772X.1966.11097123 https://www.evacranetrust.org/uploads/document/
a6e5fe111233c705b26c4a35d3f0022a54d15d9d.pdf

211   Peer, D.F. (1965). Two-queen management with package bees. *Bee-Lines* **21:** 3-7. Cited by
Crane (1966), pp.137-138, 145-146. and by Butz and Dietz (1994) loc. cit.

Peer, D.F. (1969). Two-queen management with package colonies (*Apis mellifera*). *American
Bee Journal* **109(3):**88-89. Cited by Valle, A.G.G., Guzmán-Novoa, E., Benítez, A.C. and
Rubio, J.A.Z. (2004). The effect of using two bee (*Apis mellifera* L.) queens on colony
population, honey production, and profitability in the Mexican high plateau. *Téc Pecu
Méx* **42(3):**361-377. http://cienciaspecuarias.inifap.gob.mx/index.php/Pecuarias/article/
viewFile/1404/1399

212   Walton, G.M. (1970). More honey from two-queen colonies. *New Zealand Journal of
Agriculture* **120(2):**69, 71. https://catalogue.nla.gov.au/Record/71134

Walton, G.M. (1971). Productive colony management, 1. *New Zealand Journal of
Agriculture* **122(2):**75-77. https://catalogue.nla.gov.au/Record/71134

Walton, G.M. (1972). The economics of the single-queen and two-queen systems of colony
management. New Zealand Ministry of Agriculture and Fisheries, Palmerston North, New
Zealand, 32pp.

Walton, G.M. (1974). The single-queen and two-queen systems of colony management
under commercial beekeeping conditions. *Annual Journal of Royal New Zealand Institute of
Horticulture* **No.2:**34-43.

Walsh, R.S. (1948). Two-queen colonies can increase honey production. *New Zealand
Journal of Agriculture* **77(2):**189-191. https://catalogue.nla.gov.au/Record/71134

Walsh, R.S. (1967). An experiment with queen banks. *New Zealand Beekeeper* **29(4):**14–17.

Forster, I.W. (1972). Requeening honey bee colonies without dequeening. *New Zealand
Journal of Agricultural Research* **15(2):**413-419. https://www.tandfonline.com/doi/pdf/10.108
0/00288233.1972.10421270

213   Nabors, R. (August 1, 2016) loc. cit.

214   Hesbach,W. (2016) loc. cit.

215   Heuvel, B. (March 2013), NRW, Germany. Running two-queen colonies. http://
www.beesource.com/forums/showthread.php?306234-Running-two-queen-
colonies&p=1202508#post1202508

216   Langstroth, L.L. (1853). *Langstroth on the Hive and the Honey-bee, a Bee-keeper's Manual*,
404pp. Northampton, Hopkins, Bridgman, 1853, p.234. https://babel.hathitrust.org/cgi/
pt?id=ncs1.ark:/13960/t5w671f5m&view=1up&seq=11

https://hdl.handle.net/2027/ncs1.ark:/13960/t5w671f5m

217   Moeller, F.E. (April 1976). Two queen system of honeybee colony management. Production
Research Report 161, Agricultural Research Service, United States Department of Agriculture,
15pp., Washington DC 20402. http://naldc.nal.usda.gov/download/CAT87210713/PDF

http://mesindus.ee/files/52221134-2-queen-management.pdf

218   Moeller, F.E. (1961). The relationship between colony populations and honey production as
affected by honey bee stock lines, p.8. US Department of Agriculture, Agricultural Research

Service, Production Research Report 55, 24pp. https://babel.hathitrust.org/cgi/pt?id=uc1. aa0001977776;view=1up;seq=2

Moeller, F.E. and Harp, E.R. (1965). The two-queen system simplified. *Gleanings in Bee Culture* **93(11):**679-682, 698.

Moeller, F.E. (1976) p.5 loc. cit.

219 Moeller, F.E. (1980). Managing colonies for high honey yields. *Beekeeping in the United States Apiculture Handbook* Number 335, 64-72. US Department of Agriculture. Revised October 1980. http://beesource.com/resources/usda/managing-colonies-for-high-honey-yields/ originally published in 1971 at https://thebeeyard.org/wp-content/uploads/2014/02/Beekeeping.in_.the_.United.States.pdf

[Also found online as Moeller, F.E. as Managing colonies for high honey yields at pages 23-30 in Agricultural Research Service, United States Department of Agriculture (August 1967, revised June 1971) Agriculture Handbook No. 335. *Beekeeping in the United States.* 148 pp. https://babel.hathitrust.org/cgi/pt?id=uva.x030492509&view=1up&seq=6]

Moeller, F.E. (1980). Queens, two queen system of honey production. *ABC and XYZ of Bee Culture*, pp.555-556. AI Root Co., Medina, Ohio.

220 Wade, A. (2019) loc. cit.

221 Hogg, J.A. (1981). The consolidated two-queen brood nest and queen behavior: Queens co-exist in contact through an excluder. *American Bee Journal* **121(1):**36-42. http://www.twilightmd.com/Samples/Hogg/Hogg_Halfcomb___Publications/ABJ_1981_January.pdf

Hogg, J.A. (May 1, 1983). Methods for double queening the consolidated broodnest hive: The fundamentals of queen introduction, Part I. *American Bee Journal* **123(5):**383-388. http://www.twilightmd.com/Samples/Hogg/Hogg_Halfcomb___Publications/ABJ_1983_1May.pdf

Hogg, J.A. (June 2, 1983). Methods for double queening the consolidated broodnest hive: The fundamentals of queen introduction: Part II: Conclusion. *American Bee Journal* **123(6):**450-454. http://www.twilightmd.com/Samples/Hogg/Hogg_Halfcomb___Publications/ABJ_1983_2June.pdf

Hogg, J.A. (2000). The Juniper Hill split for comb honey production. *American Bee Journal* **140(5):**368-369. http://www.twilightmd.com/Samples/Hogg/Hogg_Halfcomb___Publications/ABJ_2000_May.pdf

Hogg, J.A. (2003). The Juniper Hill plan for comb honey production. *American Bee Journal* **143(4):**285-288. http://www.twilightmd.com/Samples/Hogg/Hogg_Halfcomb___Publications/ABJ_2003_April.pdf

Hogg, J.A. (2005). The Juniper Hill plan for comb honey production, improved two-queen system. *American Bee Journal* **145(2):**138-141. http://www.twilightmd.com/Samples/Hogg/Hogg_Halfcomb___Publications/ABJ_2005_February.pdf

222 Hogg, J.A. (May 1, 1983) loc. cit.

223 Lysne, J. (1957). The two-queen system is an aid in swarm control. *Gleanings in Bee Culture* **85(4):**213, 215. https://archive.org/details/sim_gleanings-in-bee-culture_1957-04_85_4/page/212/mode/2up

Hines, T. (1960). 2 Queen system. *Gleanings in Bee Culture* **88(4):**212-213. https://archive.org/details/sim_gleanings-in-bee-culture_1960-04_88_4/page/212/mode/2up

224 Farrar, C.L. (1953) loc. cit.

225 Juniper Hill Enterprises (10 July 2012). A family video about comb honey in the halfcomb cassette: Hogg Halfcom. https://www.youtube.com/watch?v=ATjkAx1OENU

Hogg, J.A. (1980). A new comb honey concept. *American Bee Journal* **120(5):**357-362.

Betterbee (Northeast Center for Beekeeping, LLC) [US] (March 2015). Juniper Hill split. https://www.betterbee.com/images/Juniper%20Hill%20Split.pdf

Parsons, W. (1997). Using the Juniper Hill plan for comb honey production. *American Bee Journal* **137(9):**627-628.

Crane, E. (1980). New concepts in comb honey. *Bee World* **61(4):**129-130. https://www.evacranetrust.org/uploads/document/1f7cc447fc2d880ffa514507b99f2ab854543a75.pdf

226 Farrar, C.L. (1946). Two-queen colony management. United States Department of Agriculture. Agricultural Research Administration, Bureau of Entomology and Plant Quarantine, 14pp. http://ufdc.ufl.edu/AA00026061/00001/14j

https://archive.org/stream/twoquyma00unit/twoquyma00unit_djvu.txt

https://www.scribd.com/document/313855449/Two-queen-colony-managemet

http://www.immenfreunde.de/docs/2queen.pdf

https://ufdc.ufl.edu/AA00026061/00001

227 Crane, E. (1957). Second American bee journey: Part III Cuba and Mexico. *Bee World* **38(12):**301-313. https://doi.org/10.1080/000577 2X.1957.11095023 https://www.evacranetrust.org/uploads/document d8eb6b8772f4008dafb2b505b88306e080b53250.pdf https://www.evacranetrust.org/uploads/document/63b51bdc9c2af302e23c42ae73bb5a9428e7f50a.pdf https://www.evacranetrust.org/uploads/document/955e5c71044ed246bf7ba39cb86243c891027f50.pdf

228 Duff, S.R. and Furgala, B. (1990). A comparison of three non-migratory systems for managing honey bees (*Apis mellifera* L.) in Minnesota: Part I Management and productivity. *American Bee Journal* **130(1):**44-48. http://garybees.cfans.umn.edu/sites/garybees.dl.umn.edu/files/comparison_i.pdf

Duff, S.R. and Furgala, B. (1990). A comparison of three non-migratory systems for managing honey bees (*Apis mellifera* L.) in Minnesota: Part II Economic analysis. *American Bee Journal* **130(2):**121-126. https://scholar.google.com.au/scholar?hl=en&as_sdt=0%2C5&q=+Duff%2C+S.R.+and+Furgala%2C+B.+%281990b%29.++A+comparison+of+three+non-migratory+systems+for+managing+honey+bees+%28Apis+mellifera+L

229 Villarroel, D.T., Rebolledo, R.R. and Aguilera, A.P. (1998). Comparative study of honey production with one and two queens per hive in the area of Nueva Imperial, IX Region, Chile. [Estudio comparativo de producción de miel con una y dos reinas por colmena en la zona de Nueva Imperial, IX Región, Chile.] *Agro Sur* **26(2):**121-126. http://revistas.uach.cl/html/agrosur/v26n2/body/art12.htm

Rebolledo, V.R.R., Guiñez, C.G., Araneda, D.X. and Aguilera P.A. (2008). Comparative study of honey bee production with one and three queens by beehive in Neuva Imperial, IX Region, Chile. [Estudio comparativo de producción de miel con una y tres reinas por colmena en la zona de Nueva Imperial, Chile.] *Idesia* **26(2):**19-25. https://scielo.conicyt.cl/scielo.php?pid=S0718-34292008000200004&script=sci_arttext https://www.researchgate.net/publication/297710235_Comparative_study_of_honey_bee_production_with_one_and_three_queens_by_beehive_in_nueva_imperial_IX_region_Chile

Rebolledo, R.R., Riquelme, M.C., Haiquil, S., Sepúlveda, G. and Aguilera, A.P. (2011). Comparative study of honey and pollen production in a double queen system versus one queen per hive in La Araucanía Region, Chile. [Estudio comparativo de la producción de polen y miel en un sistema de doble reina versus una por colmena en La Araucanía, Chile.] *Idesia* **29:**139-144. http://www.*Idesia*.cl/Vols/2011/29-2/art18.pdf https://scielo.conicyt.cl/scielo.php?script=sci_arttext&pid=S0718-34292011000200018&lng=en&nrm=iso&tlng=en

Cengiz, E.H., Genç, F. and Cengiz, M.M. (2019). The effect of the two-queen colony management practice on colony performance and *Varroa* (*Varroa destructor*

Anderson&Trueman) infestation levels in honey bee (*Apis mellifera* L.) colonies. *Uludag Bee Journal* **19(1):**1-11. https://dergipark.org.tr/en/download/issue-full-file/45334

230　Valle, A.G.G., Guzmán-Novoa, E., Benítez, A.C. and Rubio, J.A.Z. (2004). The effect of using two bee (*Apis mellifera* L.) queens on colony population, honey production, and profitability in the Mexican high plateau. *Téc Pecu Méx* **42(3):**361-377. http://cienciaspecuarias.inifap.gob.mx/index.php/Pecuarias/article/viewFile/1404/1399

231　Victors, S. (2001). Two queen hive system from package bees. Alaska Wildflower Honey, 8pp. http://sababeekeepers.com/files/Two_Queen_Hive_System_From_Package_Bees.pdf

Winston, M. and Mitchell, M. (1986). Timing of package honey bee (Hymenoptera: Apidae) production and use of two-queen management in southwestern British Columbia, Canada. *Journal of Economic Entomology* **79(4):**952-956. https://discovery.csiro.au/primo-explore/fulldisplay?docid=TN_proquest14497293&context=PC&vid=CSIRO&lang=en_US&search_scope=All&adaptor=primo_central_multiple_fe&tab=default&query=any,contains,Winston,%20M.%20and%20Mitchell,%20M.%20(1986).%20%20Timing%20of%20package%20honey%20bee&offset=0

232　Farrar, C.L. (October 1936 and May 1946); Moeller, F.E. (April 1976); and Hogg, J.A. (1981 and May 1 and June 2, 1983) loc. cit.

Garcia, E.A.G. (1976). Contribucion Al Estudio Comparativo De La Produccion De Miel Y Cera En Colmenas Con Una Y Qos Reinas. Facultad de Medicina Veterinaria y Zootecnia (Guadalajara, Jalisco, México). 52pp. http://repositorio.cucba.udg.mx:8080/xmlui/bitstream/handle/123456789/4333/Garabito_Garcia_Estela_Alexandra.pdf?sequence=1

233　Sommerville, D. and Collins, D. (2014). Screened bottom boards. Rural Industries Research and Development Corporation Publication 14-061. 48pp. https://www.agrifutures.com.au/wp-content/uploads/publications/14-061.pdf

234　Hogg, J.A. (May 1 and June 2, 1983) loc. cit.

235　Hogg, J.A. (2000; 2003; 2005) loc. cit.

236　Farrar, C.L. (1954) loc. cit.

Appendix —
Wells Two Queen System of Beekeeping

Once in a while one comes across an apiarist who has that innate sense of how bees actually operate. They, not the bees, are a rare breed.

George Wells was such a beekeeper. His consummate knowledge of bees, and the fact that his hives were producing about twice as much honey – in the equivalent amount of bee gear – as any other Cottager beekeeper in merry old England, made him quickly famous.

Following an historic 1892 meeting with the prestigious *British Bee-Keepers' Association Quarterly Converzatione*, and the subsequent flood of correspondence that appeared in the *British Bee Journal* and the *Bee-Keepers' Record and Adviser*, Wells felt compelled to provide a personal account on how he discovered his system of beekeeping. This was to come in April 1894 when he advertised as the *"Wells System"*.

THE "WELLS SYSTEM"
Described by the Originator,
GEORGE WELLS,
AYLESFORD, KENT.
Price 6½d. Post Free.
To be had of the Author only.

This privately published booklet, long lost to the beekeeping world, has been rediscovered by *Northern Bee Books*.

Wells had already reported extensively on his method of beekeeping – running two queens in separate hives divided by a thin punched division board. These hives shared common honey supers over the active bee season reverting back to separate single-queen hives – still in a doubled hive setup – from late autumn. His scheme and the details of its set up had been widely canvassed in the *British Bee Journal* (BBJ) though the intricacies of how he managed his bees day-to-day and across the seasons were less well known. This publication reveals these secrets of working bees.

With his untimely death in 1908, the Wells System of beekeeping fell into disuse and by the mid 1920s his scheme was all but forgotten. The key reasons for his system being abandoned were the labour required to fine tune the conditions necessary for doubled hives to operate successfully and the bespoke equipment needed. Less capable beekeepers of his day were unable to replicate his success despite the many kits for his hive being advertised for sale in the BBJ.

But good ideas never die. The Wells System of beekeeping is now widely practised using standardised bee equipment. The modern doubled hive system promoted by William Hesbach and Ray Nabors over the past decade have much to offer but lack of the subtlety of the original Wells System. Happily, however, the reinvented scheme has proved popular amongst sideline beekeepers.

It is now time to revisit George Wells as one of the great contributors to our understanding of the inner workings of the honey bee colony. Wells detailed account of swarm control and his ever only working with young and productive queens is a tribute to a great man with a clear vision of how to keep, rather than, own bees. And his doubled hive set up, now using more conventional gear, certainly works well and should be a magnet to the beekeeper owning only a handful of hives.

Alan Wade

Canberra Region Beekeepers
September 2021

~Guide Book Pamphlet~

ON THE

TWO QUEEN SYSTEM

OF

BEE KEEPING

BY

G. WELLS,

AYLESFORD,

NEAR MAIDSTONE,

KENT.

Price 6½d., Post Free, in the United Kingdom.

Published by G. F. GAY,

"SNODLAND STEAM PRINTING WORKS,"

MALLING ROAD.

[ENTERED AT STATIONERS' HALL.]

PREFACE.

IT was early in 1892 when I first made the Two Queen System known to the public, and, at that time, I quite thought that by what I published in the B.B. Journal and other Publications, and with the assistance of our esteemed Editors of the B.B. Journal, that nothing further would be required from me upon the subject, but as time went on enquiries became more and more numerous and at times quite wearisome, and I still thought that the numerous private letters I have written, and by what has been published from time to time in various publications, that these enquiries would become unnecessary, but such is not the case, as I still receive large numbers of enquiries both from this and other countries pressing me for a Pamphlet upon the subject, and I have no other excuse to offer for attempting to supyly this want.

I have not gone out of my way to bring in anything but what I have proved in practice, and that I have endeavoured to make as plain as I could, but I do not pretend to think that I have perfected everything in the system of Bee keeping, but quite to the contrary, for I seem to think that there is much more to be learnt than is already known about that wonderful little insect the "Honey Bee," and in publishing this Pamphlet my desire is to assist Bee keepers as far as possible to enable them to stand on a firm footing with the Foreign Honey supply, and this I fear will be a hard matter to do, unless something is done to increase the output far beyond what can be expected on an average from hives with but one Queen in each, of course I have mentioned many things which have been written by able and efficient writers upon Bee keeping, and I beg to apologise to those gentlemen, and to say that I could not write my own experience without it, as I have learnt all I know upon the subject of Bee keeping from what I have read from other publications, coupled with my own experience in practice.

G. WELLS,
AYLESFORD,
NEAR MAIDSTONE,
KENT,
ENGLAND.

INTRODUCTION.

PREVIOUS to the year of 1890 I had been in the habit of preserving all my surplus young Queens through the Winter in case any Stock should be found Queenless in the following Spring. My method of preserving them was by putting two Queens with their Bees in one Hive, dividing the two lots by inserting a fine wire division in the centre of the Hive, having one Queen with her Bees on each side, being very careful that there was no way in which either Queen or Bees could come in actual contact with each other, unless they go first out of the entrance to their own compartment and then into the entrance of the neighbouring compartment, and to avoid this a temporary division was placed in the centre of the flight board, extending close up under the porch and to the outside edge, with a small piece of wood to divide the entrance between the front inside edge of the Hive and the division on the flight board. Now in the following Spring if any Stocks were found Queenless, one of these spare Queens was taken and introduced into the Queenless Stock, and then the wire division was withdrawn and the remaining Queen allowed to have all the Bees and Brood which had been previously kept from her as long as the other Queen was in the Hive, and I never saw any signs whatever of any fighting or discontent when these two lots of Bees were thus allowed to mix together and so form one lot instead of two as it had been. If I had yet any Queens to spare after I had supplied all the Queenless Stocks with young Queens, the wire division and the two Queens were allowed to remain in the Hives until they required Supering, and then, if I did not know of anyone who would be glad of the Surplus Queens I had to spare, the most backward one (all things considered) was caught and killed, and then withdraw the wire division and allow all the Bees to mix and form one lot.

Thus I went on from year to year until 1890 when I had but one Hive with two Queens in it, and a friend of mine wanted me to save the spare one for a short time for him and he delayed fetching it. Now these two Queens generated Brood so rapidly that I was compelled to withdraw frame after frame of hatching Brood and replace them with empty combs to prevent them swarming. The frames of Brood were given to the other Hives, which had but one Queen each, to strengthen them until the beginning of June, and at that time all the Hives were full of Bees and Supered. Yet the Hive with the two Queens was boiling over with Bees and Combs full of Brood. Then, for the first time, the thought struck me that if the Bees would mix together peaceably, by killing one Queen and withdrawing the wire division, why should they not work as well in the Supers from the two Queens as though there was but one Queen. So I placed a sheet of Queen Excluder Zinc on the top of the frames, making quite sure that the two Queens could not come in actual contact, and put on the Supers. At this time some of my other Hives required the second crate of Supers and notwithstanding that advantage, together with the disadvantage, the Hive with the two Queens had been labouring under, I had the pleasure of taking very much more Surplus Honey from the Hive with the two Queens in it than I took from any other. I did not keep any account of the difference, but it was so surprising to me that I thought I should never have a Hive again with but one Queen in it, and so in the Autumn of the same year I packed all my Hives up for the Winter with two Queens in each. But previous to this I had noticed that in using the wire net for a division it caused the two lots of Bees to cluster as far from it as they could, and only filled the space next to the wire division late in the Spring when they were compelled to do so for the want of room, and so my mind had been wondering what I could get for a division that would encourage the Bees to cluster upon it on both sides in the Winter as well as in the Summer. My mind fell upon the thin soft wood (Best Yellow Pine), first making holes through it with a small brad-awl, about half an inch apart, thus :—

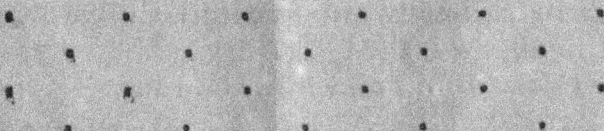

    •   •   •   •   •   •

     •   •   •   •   •

    •   •   •   •   •   •

And then burn them through with an hot wire, one eighth of an inch thick, then binding it with rather light tin to prevent it warping. Thus each of my Hives were supplied for the first time with one of these thin soft wood perforated division boards, just the same as I use now in the place of the wire net divisions which I had previously used, and to my great satisfaction I found that the Bees in every case clustered close up to and upon them when they were first put into the Hive and continued to do so right through the Winter and but very few holes in the division boards were propolised up, and in the Spring of 1891 the Bees were in much better condition than I had ever had them before and they all generated Brood so rapidly that for the first time I had them in condition to store some Surplus Honey in the Supers from the early fruit bloom.

In the Autumn of 1890 I had eleven Hives packed for Winter with two Queens in each of them and an abundance of sealed stores. They were not disturbed until the middle of March when I found that all were in good condition with lots of young Bees and each Queen had much more Brood than I had ever seen before at this time of the year. They had also abundance of food, showing that two lots in one Hive, as described, do not consume so much food during the Winter in proportion as single stocks do, and thus I escaped the trouble and expense of Spring feeding.

These Eleven Stocks gave me during the Season :—

	£	s.	d.
312 one lb. Sections at 1/-	15	12	0
1069 lbs. of Extracted Honey at 9d. ...	40	1	9
40½ lbs. of Wax at 2/-...	4	1	0
Total from Eleven Hives... ...	59	14	9
Deduct Total Expenditure during the Year	8	9	8
Balance for Labour	£51	5	1

The above figures give an average of 125½ lbs. of Honey per Hive, and an average of over 3½ lbs. of Wax per Hive. After these results I quite made up my mind never to keep Bees again with less than two Queens in each Hive, and at the same time I decided to limit my number of Hives to ten. I always preserve all my surplus young Queens through the Winter, but reduce the number to ten hives in the Spring.

I had many friends to see my Bees in the Spring of 1892, and several suggested that I ought to run one or two Hives with but one Queen in them, and thus prove the difference between the two systems, and, as I had ten Hives, I thought the best proof would be to work five Hives with but one Queen in each of them, and five Hives with two Queens in each of them, so this plan was adopted, and the results were as follows :—the Five Single Queen Stocks gave 205 lbs. of Surplus Honey, an average of just 41 lbs. each; the Five Double Queen Stocks gave 789 lbs. of Surplus Honey, an average of very nearly 158 lbs. each.

I think these results make it very plain that it pays very much better to work with two Queens in a Hive than it does to work with but one Queen in a Hive.

The financial position of the Bees for 1892 is as follows:—

	£	s.	d.
940 lbs of Extracted Honey at an average of 9d. per lb.	35	5	0
54 one lb. Sections of Comb Honey at 1/-	2	14	0
30 lbs. of Wax at 2/-	3	0	0
	40	19	0
Deduct Total Expenditure during the Year	5	1	10
Balance for Labour	£35	17	2

In the Autumn of 1892 all Hives were again put into Winter quarters with two Queens in each and abundance of stores. As usual all went through the Winter well and opened in grand condition about the middle of March. In 1893 I again had many friends to look at the Bees, so tha they were opened and taken out of their Hives a very great deal more than was good for them, and early in May I discovered that one Queen was missing from one compartment of one Hive and nearly all the Bees had left the Queenless side and crowded into the other side which still retained a Queen, and knowing this Queen to be a very superior one I decided to run it for young Queens, so that my Surplus Honey in 1893 was taken from nine Stocks, with two Queens in each, and the results were as follows :—1,223 lbs., an average of very nearly 136 lbs. per Hive.

The financial position of the Bees for 1893 is as follows:—

	£	s.	d.
1115 lbs. of Extracted Honey at an average of 9d. per lb.	41	16	3
108 one lb. Sections of Comb Honey at 1/-	5	8	0
19 lbs. of Wax at 2/-	1	18	0
	49	2	3
Deduct Total Expenditure during the Year	1	0	9
Balance for Labour	£48	1	6

Now the total amount of Surplus Honey taken from all the Two-Queen Hives in the three years amounts to 3,393 lbs., or an average of a trifle under 136 lbs. per Hive per year. In addition we have to add 81½ lbs. of Bees Wax, an average of a little over 3lbs. per Hive per year.

The financial position of the Two-Queen Stocks for the three years is as follows :—

	£	s.	d.
Extracted Honey, 2,946 lbs. at 9d. ...	110	9	6
Comb Honey, 447 lbs. at 1/-	22	7	0
Bees Wax, 81½ lbs. at 2/-	8	3	0
	140	19	6
Deduct Total Expenditure in the Three Years	14	12	3
Balance for Labour	£126	7	3

Or an average of £5/1/1 per Hive per year, for three successive years.

These prices were obtained after the Honey, etc., was sold, but those which have previously appeared in print were but an estimate before the Honey, etc., was sold, my own consumption also being credited at the same rate.

HOW IT HAS BEEN DONE.

TO get the full advantage of the system we must begin in the Autumn by selecting a Hive to hold 20 standard size Frames, divided in the centre with the thin soft wood perforated division board (any kind of metal used as a division would prove fatal to the system), take two young fertile and prolific Queens and place one on each side of the division, with Bees enough to well cover at least seven Frames, and there should be enough sealed stores in these seven Frames to cover about three-fourths of the surface of the same. If the Bees on these seven frames, on each side of the division, were rather crowded, so much the better. The empty space at either end of the Hive should be filled up with something warm.

I use for this purpose some thick dummies, made by nailing two thin boards, 14¾ inches by 8⅜ inches by ⅜ inch, on two strips of wood, a thin bottom (wood), and the top the same thickness as the top bars of Frames, and they hang in the Hive in the same manner as the Frames. The two strips which the boards are nailed to are placed about two inches from the ends, the ends proper are formed by tacking a piece of unbleached calico over the ends of the dummy, and the hollow space between the two boards is filled up with oat chaff through a slot made along the middle of the top bar, the chaff bulges the calico and thus it fits tight to the sides of the Hive. I have some to fill the space of one comb, some two combs, and some three combs.

I lower my floor boards three inches in front and nothing at the back, giving a sloping floor with space below the combs, having a wood block to fill the large space, with an entrance four inches wide, and thus they remain until about the middle of the following March, when they are examined, and note the condition of each, if any of the combs are not then covered with Bees they are removed and the space filled with dummies. If any are getting short of stores, an empty frame is removed and a Frame of sealed stores is put in its place, first made to about the same temperature of the Hive.

Now the floor boards are taken out, scraped, and washed with strong soap and soda water (hot), and rinsed with clean hot water with a little Carbolic Acid added to it. The floor board is then rubbed dry, and three half-lumps of Napthaline is laid on it, then replace it and fasten it close up to the bottom of the Hive, leaving only about the space of $\frac{3}{8}$ in. between the bottom bars of Frames and the floor board.

The entrance is closed, only allowing the Bees just room enough to crowd in and out, gradually opening it as they increase in number.

Now the Bees will soon be somewhat crowded in the Hive, and as soon as that is so, remove a dummy and slip in another comb, with or without stores as the inside conditions require, but always let these combs be made as warm (right through them) as the Hive, and do not disturb the Bees more than can be avoided. Cover up as quickly and as warm as possible, and in a few days another comb will be wanted which should be added in the same way, but the Bees should always be in a condition to take possession of the extra combs directly they are given, and thus they should be led on until the Brood Nests are boiling with Bees.

Almost any Hive can be used for the purpose. Even a ten-frame Hive would do by placing the perforated division board in the centre, giving five combs on each side. But in this case a second storey is necessary in order to give each Queen room for the Brood Nests, and for this purpose each Queen should have ten Standard size Frames. Those who have Combination Hives of fourteen and fifteen Frames could convert them for the System by cutting the second entrance, either at the opposite end or at the side. To give the Queens room, a crate of Shallow Frames could be added on the top of the Standard Frames, but there must be the second division board used in this Crate of Shallow or other Frames. It must be placed exactly parallel with the one in the lower body, and it must be proof against the Bees passing from one side to the other. But a Hive that will hold twenty Standard size Frames in the lower body is much to be preferred.

There are now many different kinds of Hives of this size (and which have been designed expressly for the two Queen System) in the Market. Of course I have my fancy, but I will not pretend to say which is the best for the purpose, as I have not had the pleasure of seeing all of them, and a great deal depends upon a person's fancy; but whatever Hive is used there must be a sheet of Queen Excluder Zinc used, large enough to cover the whole of the top bars of the Frames before the Supers are put on, and if the Zinc is mounted on $\frac{1}{4}$ inch strips (as some are), one of these strips must be so placed, so that when the Zinc is in position, the strips run right along over the division board below; but it is best to have the Zinc in two parts and let them meet in the middle of the Hive, so that the joint is right over the division board, in which case, one Brood Nest can be examined without disturbing the other, making quite sure that neither of the Queens can get past the Excluder Zinc. In the case of a Hive holding twenty Frames in the lower body it would be rather handiest if the Super crates were in two parts (but only for the convenience of handling) in which case there must be a Bee Space left under the bottom edges of both crates where they meet in the middle, so that all the worker Bees from both Brood Nests can get to any and all the Supers, or in other words, it is one crate to the Bees, but two crates to the Bee Keeper, and in no case should the Queen Excluder Zinc and the Supers be put on until it is quite necessary in order to give the Bees room. There is much more mischief done by far in giving the Bees Surplus Chamber too soon, than there is in not giving them room soon enough, even if it causes them to swarm. When it is necessary to give more room they should have but one crate at first, and if there is not Bees enough to take possession of a whole crate of sections, do not let them have a whole one, but such a part of a crate that will be filled with Bees at once, and so add more sections as fast as they are required to give the Bees room. This first crate should be placed in the centre of the Hive so that the Bees from both compartments can get into it. The two ends of the Excluder Zinc which are not covered with the section crate must be quilted over to keep the Bees in, and any amount of warm material may be wrapped round and over the section crate, and by the time this first crate is well crowded with Bees, a second whole crate may be given. If one side of the Hive is stronger than the other, set the crate which was first put on over the weakest side, and the empty one close up to it; those two ends where the Bee space is left coming together in the centre of the Hive.

If shallow Frames are used they should be given in the

same way, one or two combs at a time, filling the empty space with the thick dummies as previously described; by the time these two crates are well filled with Bees and about half full of Honey, they can have two more empty crates added.

I prefer to remove the first two crates and place the two empty ones underneath, and be sure and keep them always well packed up so that no heat is lost, and when the top two crates are finished, they can be removed, and two other empty ones put next to the Excluder Zinc as before, or they may be allowed to remain on the Hive until the end of the season.

In using shallow Frames for extracted Honey, I have had them so packed up one upon the other to the height of six stories, in which case the Honey is well ripened, and more Wax will be got from the Cappings.

If one thinks it is too much trouble to remove the surplus crates when they are half filled with Honey in order to put the empty ones next to the Queen Excluder Zinc, the empty ones may be put on the top, in which case the Combs or Sections can be given a few at a time. I have had good results both ways, but I prefer giving the empty crates, either Sections or shallow Frames, next to the Queen Excluder Zinc.

I have received many letters from Bee Keepers complaining that they could not get the Bees to take readily to the Supers, this is not the fault of the Bees, but that of the Bee Keeper in giving the Supers too soon, or giving too much extra room in the Supers at one time.

If the foregoing instructions are strictly adhered to, there will be no difficulty whatever in that respect.

SWARMING.

I HAVE not been able to effectually prevent swarming until 1893. In that year I had but one swarm
and that one I crowded the Bees on purpose to make
them swarm, and the peculiarity of the season may
have had something to do with the non-swarming of
the other Stocks.

In this system of Bee Keeping it is best to prevent
swarming from those Stocks which are being worked
for Surplus Honey, by carefully watching and giving
more room little by little just as they require it, and
if some of them should swarm they may be turned to
good account. One very great objection to swarming
is that when the Bees on one side of the division prepare for swarming, the other is nearly sure to do the
same, and when the swarm comes off, both Queens
leave with the Bees. Therefore Nucleus Hives should
always be in readiness during the swarming season.
If a swarm comes off and someone is present at the
time, they should be closely watched and as soon as
it is seen the exact spot where they intend to cluster,
go to that spot and watch for one of the Queens to
settle, and if she is seen she should be captured. The
person should have something in his pocket at the
time to put one of the Queens in (I use an empty
match box), and as soon as the one Queen is captured
and in the box she may be put into the pocket or in

any warm dark place, and if a few of her Bees were with her so much the better. Now leave the Bees to cluster with the one Queen and get your Nucleus Hives close up to the Hive where the swarm came from; now remove all the surplus crates from the Hive and stand them aside so that the Bees can neither get out of them nor into them; now examine both Brood Nests and take a comb with a good Queen cell, cutting out all but the one. Put this comb and adhering Bees into the centre of one Nucleus Hive and another comb of hatching Brood next to the side of the comb on which the Queen cell is attached; now a frame of stores on the other side of the comb which contains the Queen cell; now brush enough young Bees from other combs to well cover the three combs in Nucleus Hive. Pack close with dummies and cover up warmly and proceed with as many more Nucleus Hives as you have Queen cells and Bees to form them. Now fill both Brood Nests up in parent Hive with clean empty combs or full sheets of comb foundation, placing those combs which contain Brood (no Queen cells must be left on them) close up to either side of the perforated division board; now put on the Excluder Zinc and replace all surplus crates just as they were before the swarm came off. The swarm should have been shaded as soon as they had clustered. Now take the Queen out of the match box and let her run in at one of the entrances (no matter which); now stop that entrance up so that the Bees can neither get in or out; now lower the floor board in front of the other entrance and lay a large board or some substitute on the flight board and the other end on the ground. Cover this with a large clean cloth. Take a large skip and another cloth, go to your swarm and with a sudden jerk eject the Bees into the skip, throw the cloth loosely over the skip and Bees, and carry them to the Hive they came from. Lay a stone or a piece of wood in front of the Hive on the cloth which covers the board; now turn your skip and let the Bees fall gently on to the cloth and let the edge of the skip rest on the stone and rest yourself for a minute; now raise the skip and jar out a few Bees close up to the mouth of the parent Hive, and as soon as some of them have run into the Hive, jar out more, and when they are well on the run into the Hive, jar out the remainder. As soon as they have nearly all run into the Hive, the other side can be opened and the floor board, where the swarm was run in, can be raised and fastened up in its usual position, as soon as the Bees are sufficiently clear so as not to crush any of them.

I find Bees treated in this way go to work at once with a will, and as to storing honey, they will be but very little, if any, behind those which did not swarm. All the Nucleus swarms should be fed with a slow feeder, and kept in as warm a position as possible. These young Queens in Nucleus Hives are very valuable in the Autumn to replace those which have gone through two full seasons. If possible none should be kept after two full seasons.

In conclusion I would say—Well crowd the Bees in the Autumn on abundance of sealed stores and cover them up as warmly as possible. Give room below the combs with an entrance of not less than 4 inches wide by $\frac{3}{8}$ inch deep. Do not disturb the Bees before the middle of March, and then only on a warm day. Cover up as quickly and as warmly as possible. Clean floor boards and reduce space below combs to from $\frac{3}{8}$ inch to $\frac{1}{2}$ inch deep. Give small entrance, and add an extra comb when the Bees must have more room, by slipping it in on the outside, with or without stores in it, as the condition of the Hive requires. Do not disturb the Bees more than can possibly be avoided at any time.

I have now only to say that I have tried to make my system of working Bees as plain as I can, and I hope it will be understood. I also hope that all who try the system will be successful; but persons must not expect the best results from the system unless they have a fair knowledge of Bee keeping, and are willing to carry out the instructions until something better is discovered.

In consequence of the difficulty experienced by Bee Keepers in obtaining division boards (perforated), I have now made arrangements to supply them the same as I use, at 2/6 each, post free.

G. F. Gay, Steam Printer, Snodland.

www.ingramcontent.com/pod-product-compliance
Lightning Source LLC
Chambersburg PA
CBHW050037220326
41599CB00040B/7195